Circadian Rhythms: A Very Short Introduction

VERY SHORT INTRODUCTIONS are for anyone wanting a stimulating and accessible way into a new subject. They are written by experts, and have been translated into more than 45 different languages.

The series began in 1995, and now covers a wide variety of topics in every discipline. The VSI library now contains over 500 volumes—a Very Short Introduction to everything from Psychology and Philosophy of Science to American History and Relativity—and continues to grow in every subject area.

Very Short Introductions available now:

ACCOUNTING Christopher Nobes
ADOLESCENCE Peter K. Smith
ADVERTISING Winston Fletcher
AFRICAN AMERICAN RELIGION
    Eddie S. Glaude Jr
AFRICAN HISTORY John Parker and
    Richard Rathbone
AFRICAN RELIGIONS Jacob K. Olupona
AGEING Nancy A. Pachana
AGNOSTICISM Robin Le Poidevin
AGRICULTURE Paul Brassley and
    Richard Soffe
ALEXANDER THE GREAT
    Hugh Bowden
ALGEBRA Peter M. Higgins
AMERICAN HISTORY Paul S. Boyer
AMERICAN IMMIGRATION
    David A. Gerber
AMERICAN LEGAL HISTORY
    G. Edward White
AMERICAN POLITICAL HISTORY
    Donald Critchlow
AMERICAN POLITICAL PARTIES
    AND ELECTIONS L. Sandy Maisel
AMERICAN POLITICS
    Richard M. Valelly
THE AMERICAN PRESIDENCY
    Charles O. Jones
THE AMERICAN REVOLUTION
    Robert J. Allison
AMERICAN SLAVERY
    Heather Andrea Williams
THE AMERICAN WEST Stephen Aron
AMERICAN WOMEN'S HISTORY
    Susan Ware

ANAESTHESIA Aidan O'Donnell
ANARCHISM Colin Ward
ANCIENT ASSYRIA Karen Radner
ANCIENT EGYPT Ian Shaw
ANCIENT EGYPTIAN ART AND
    ARCHITECTURE Christina Riggs
ANCIENT GREECE Paul Cartledge
THE ANCIENT NEAR EAST
    Amanda H. Podany
ANCIENT PHILOSOPHY Julia Annas
ANCIENT WARFARE Harry Sidebottom
ANGELS David Albert Jones
ANGLICANISM Mark Chapman
THE ANGLO-SAXON AGE John Blair
ANIMAL BEHAVIOUR
    Tristram D. Wyatt
THE ANIMAL KINGDOM
    Peter Holland
ANIMAL RIGHTS David DeGrazia
THE ANTARCTIC Klaus Dodds
ANTISEMITISM Steven Beller
ANXIETY Daniel Freeman and
    Jason Freeman
THE APOCRYPHAL GOSPELS
    Paul Foster
ARCHAEOLOGY Paul Bahn
ARCHITECTURE Andrew Ballantyne
ARISTOCRACY William Doyle
ARISTOTLE Jonathan Barnes
ART HISTORY Dana Arnold
ART THEORY Cynthia Freeland
ASIAN AMERICAN HISTORY
    Madeline Y. Hsu
ASTROBIOLOGY David C. Catling
ASTROPHYSICS James Binney

ATHEISM Julian Baggini
AUGUSTINE Henry Chadwick
AUSTRALIA Kenneth Morgan
AUTISM Uta Frith
THE AVANT GARDE David Cottington
THE AZTECS David Carrasco
BABYLONIA Trevor Bryce
BACTERIA Sebastian G. B. Amyes
BANKING John Goddard and
    John O. S. Wilson
BARTHES Jonathan Culler
THE BEATS David Sterritt
BEAUTY Roger Scruton
BEHAVIOURAL ECONOMICS
    Michelle Baddeley
BESTSELLERS John Sutherland
THE BIBLE John Riches
BIBLICAL ARCHAEOLOGY Eric H. Cline
BIOGRAPHY Hermione Lee
BLACK HOLES Katherine Blundell
BLOOD Chris Cooper
THE BLUES Elijah Wald
THE BODY Chris Shilling
THE BOOK OF MORMON
    Terryl Givens
BORDERS Alexander C. Diener and
    Joshua Hagen
THE BRAIN Michael O'Shea
THE BRICS Andrew F. Cooper
THE BRITISH CONSTITUTION
    Martin Loughlin
THE BRITISH EMPIRE Ashley Jackson
BRITISH POLITICS Anthony Wright
BUDDHA Michael Carrithers
BUDDHISM Damien Keown
BUDDHIST ETHICS Damien Keown
BYZANTIUM Peter Sarris
CALVINISM Jon Balserak
CANCER Nicholas James
CAPITALISM James Fulcher
CATHOLICISM Gerald O'Collins
CAUSATION Stephen Mumford and
    Rani Lill Anjum
THE CELL Terence Allen and
    Graham Cowling
THE CELTS BarryCunliffe
CHAOS Leonard Smith
CHEMISTRY Peter Atkins
CHILD PSYCHOLOGY Usha Goswami
CHILDREN'S LITERATURE
    Kimberley Reynolds

CHINESE LITERATURE Sabina Knight
CHOICE THEORY Michael Allingham
CHRISTIAN ART Beth Williamson
CHRISTIAN ETHICS D. Stephen Long
CHRISTIANITY Linda Woodhead
CIRCADIAN RHYTHMS
    Russell G. Foster and Leon Kreitzman
CITIZENSHIP Richard Bellamy
CIVIL ENGINEERING David Muir Wood
CLASSICAL LITERATURE William Allan
CLASSICAL MYTHOLOGY
    Helen Morales
CLASSICS Mary Beard and John Henderson
CLAUSEWITZ Michael Howard
CLIMATE Mark Maslin
CLIMATE CHANGE Mark Maslin
COGNITIVE NEUROSCIENCE
    Richard Passingham
THE COLD WAR Robert McMahon
COLONIAL AMERICA Alan Taylor
COLONIAL LATIN AMERICAN
    LITERATURE Rolena Adorno
COMBINATORICS Robin Wilson
COMEDY Matthew Bevis
COMMUNISM Leslie Holmes
COMPLEXITY John H. Holland
THE COMPUTER Darrel Ince
COMPUTER SCIENCE Subrata Dasgupta
CONFUCIANISM Daniel K. Gardner
THE CONQUISTADORS
    Matthew Restall and
    Felipe Fernández-Armesto
CONSCIENCE Paul Strohm
CONSCIOUSNESS Susan Blackmore
CONTEMPORARY ART
    Julian Stallabrass
CONTEMPORARY FICTION
    Robert Eaglestone
CONTINENTAL PHILOSOPHY
    Simon Critchley
COPERNICUS Owen Gingerich
CORAL REEFS Charles Sheppard
CORPORATE SOCIAL
    RESPONSIBILITY Jeremy Moon
CORRUPTION Leslie Holmes
COSMOLOGY Peter Coles
CRIME FICTION Richard Bradford
CRIMINAL JUSTICE Julian V. Roberts
CRITICAL THEORY
    Stephen Eric Bronner
THE CRUSADES Christopher Tyerman

CRYPTOGRAPHY Fred Piper and
    Sean Murphy
CRYSTALLOGRAPHY A. M. Glazer
THE CULTURAL REVOLUTION
    Richard Curt Kraus
DADA AND SURREALISM
    David Hopkins
DANTE Peter Hainsworth and
    David Robey
DARWIN Jonathan Howard
THE DEAD SEA SCROLLS Timothy Lim
DECOLONIZATION Dane Kennedy
DEMOCRACY Bernard Crick
DEPRESSION Jan Scott and
    Mary Jane Tacchi
DERRIDA Simon Glendinning
DESCARTES Tom Sorell
DESERTS Nick Middleton
DESIGN John Heskett
DEVELOPMENTAL BIOLOGY
    Lewis Wolpert
THE DEVIL Darren Oldridge
DIASPORA Kevin Kenny
DICTIONARIES Lynda Mugglestone
DINOSAURS David Norman
DIPLOMACY Joseph M. Siracusa
DOCUMENTARY FILM
    Patricia Aufderheide
DREAMING J. Allan Hobson
DRUGS Les Iversen
DRUIDS Barry Cunliffe
EARLY MUSIC Thomas Forrest Kelly
THE EARTH Martin Redfern
EARTH SYSTEM SCIENCE Tim Lenton
ECONOMICS Partha Dasgupta
EDUCATION Gary Thomas
EGYPTIAN MYTH Geraldine Pinch
EIGHTEENTH-CENTURY BRITAIN
    Paul Langford
THE ELEMENTS Philip Ball
EMOTION Dylan Evans
EMPIRE Stephen Howe
ENGELS Terrell Carver
ENGINEERING David Blockley
ENGLISH LITERATURE Jonathan Bate
THE ENLIGHTENMENT John Robertson
ENTREPRENEURSHIP Paul Westhead
    and Mike Wright
ENVIRONMENTAL ECONOMICS
    Stephen Smith

ENVIRONMENTAL POLITICS
    Andrew Dobson
EPICUREANISM Catherine Wilson
EPIDEMIOLOGY Rodolfo Saracci
ETHICS Simon Blackburn
ETHNOMUSICOLOGY Timothy Rice
THE ETRUSCANS Christopher Smith
EUGENICS Philippa Levine
THE EUROPEAN UNION John Pinder
    and Simon Usherwood
EVOLUTION Brian and
    Deborah Charlesworth
EXISTENTIALISM Thomas Flynn
EXPLORATION Stewart A. Weaver
THE EYE Michael Land
FAMILY LAW Jonathan Herring
FASCISM Kevin Passmore
FASHION Rebecca Arnold
FEMINISM Margaret Walters
FILM Michael Wood
FILM MUSIC Kathryn Kalinak
THE FIRST WORLD WAR
    Michael Howard
FOLK MUSIC Mark Slobin
FOOD John Krebs
FORENSIC PSYCHOLOGY David Canter
FORENSIC SCIENCE Jim Fraser
FORESTS Jaboury Ghazoul
FOSSILS Keith Thomson
FOUCAULT Gary Gutting
THE FOUNDING FATHERS
    R. B. Bernstein
FRACTALS Kenneth Falconer
FREE SPEECH Nigel Warburton
FREE WILL Thomas Pink
FRENCH LITERATURE John D. Lyons
THE FRENCH REVOLUTION
    William Doyle
FREUD Anthony Storr
FUNDAMENTALISM Malise Ruthven
FUNGI Nicholas P. Money
THE FUTURE Jennifer M. Gidley
GALAXIES John Gribbin
GALILEO Stillman Drake
GAME THEORY Ken Binmore
GANDHI Bhikhu Parekh
GENES Jonathan Slack
GENIUS Andrew Robinson
GEOGRAPHY John Matthews and
    David Herbert

GEOPOLITICS Klaus Dodds
GERMAN LITERATURE Nicholas Boyle
GERMAN PHILOSOPHY Andrew Bowie
GLOBAL CATASTROPHES Bill McGuire
GLOBAL ECONOMIC HISTORY
    Robert C. Allen
GLOBALIZATION Manfred Steger
GOD John Bowker
GOETHE Ritchie Robertson
THE GOTHIC Nick Groom
GOVERNANCE Mark Bevir
GRAVITY Timothy Clifton
THE GREAT DEPRESSION AND THE
    NEW DEAL Eric Rauchway
HABERMAS James Gordon Finlayson
THE HABSBURG EMPIRE
    Martyn Rady
HAPPINESS Daniel M. Haybron
THE HARLEM RENAISSANCE
    Cheryl A. Wall
THE HEBREW BIBLE AS LITERATURE
    Tod Linafelt
HEGEL Peter Singer
HEIDEGGER Michael Inwood
HERMENEUTICS Jens Zimmermann
HERODOTUS Jennifer T. Roberts
HIEROGLYPHS Penelope Wilson
HINDUISM Kim Knott
HISTORY John H. Arnold
THE HISTORY OF ASTRONOMY
    Michael Hoskin
THE HISTORY OF CHEMISTRY
    William H. Brock
THE HISTORY OF LIFE Michael Benton
THE HISTORY OF MATHEMATICS
    Jacqueline Stedall
THE HISTORY OF MEDICINE
    William Bynum
THE HISTORY OF TIME
    Leofranc Holford-Strevens
HIV AND AIDS Alan Whiteside
HOBBES Richard Tuck
HOLLYWOOD Peter Decherney
HOME Michael Allen Fox
HORMONES Martin Luck
HUMAN ANATOMY Leslie Klenerman
HUMAN EVOLUTION Bernard Wood
HUMAN RIGHTS Andrew Clapham
HUMANISM Stephen Law
HUME A. J. Ayer

HUMOUR Noël Carroll
THE ICE AGE Jamie Woodward
IDEOLOGY Michael Freeden
INDIAN CINEMA Ashish Rajadhyaksha
INDIAN PHILOSOPHY Sue Hamilton
THE INDUSTRIAL REVOLUTION
    Robert C. Allen
INFECTIOUS DISEASE Marta L. Wayne
    and Benjamin M. Bolker
INFORMATION Luciano Floridi
INNOVATION Mark Dodgson and
    David Gann
INTELLIGENCE Ian J. Deary
INTELLECTUAL PROPERTY
    Siva Vaidhyanathan
INTERNATIONAL LAW Vaughan Lowe
INTERNATIONAL MIGRATION
    Khalid Koser
INTERNATIONAL RELATIONS
    Paul Wilkinson
INTERNATIONAL SECURITY
    Christopher S. Browning
IRAN Ali M. Ansari
ISLAM Malise Ruthven
ISLAMIC HISTORY Adam Silverstein
ISOTOPES Rob Ellam
ITALIAN LITERATURE
    Peter Hainsworth and David Robey
JESUS Richard Bauckham
JOURNALISM Ian Hargreaves
JUDAISM Norman Solomon
JUNG Anthony Stevens
KABBALAH Joseph Dan
KAFKA Ritchie Robertson
KANT Roger Scruton
KEYNES Robert Skidelsky
KIERKEGAARD Patrick Gardiner
KNOWLEDGE Jennifer Nagel
THE KORAN Michael Cook
LANDSCAPE ARCHITECTURE
    Ian H. Thompson
LANDSCAPES AND
    GEOMORPHOLOGY
    Andrew Goudie and Heather Viles
LANGUAGES Stephen R. Anderson
LATE ANTIQUITY Gillian Clark
LAW Raymond Wacks
THE LAWS OF THERMODYNAMICS
    Peter Atkins
LEADERSHIP Keith Grint

LEARNING Mark Haselgrove
LEIBNIZ Maria Rosa Antognazza
LIBERALISM Michael Freeden
LIGHT Ian Walmsley
LINCOLN Allen C. Guelzo
LINGUISTICS Peter Matthews
LITERARY THEORY Jonathan Culler
LOCKE John Dunn
LOGIC Graham Priest
LOVE Ronald de Sousa
MACHIAVELLI Quentin Skinner
MADNESS Andrew Scull
MAGIC Owen Davies
MAGNA CARTA Nicholas Vincent
MAGNETISM Stephen Blundell
MALTHUS Donald Winch
MANAGEMENT John Hendry
MAO Delia Davin
MARINE BIOLOGY Philip V. Mladenov
THE MARQUIS DE SADE John Phillips
MARTIN LUTHER Scott H. Hendrix
MARTYRDOM Jolyon Mitchell
MARX Peter Singer
MATERIALS Christopher Hall
MATHEMATICS Timothy Gowers
THE MEANING OF LIFE Terry Eagleton
MEASUREMENT David Hand
MEDICAL ETHICS Tony Hope
MEDICAL LAW Charles Foster
MEDIEVAL BRITAIN John Gillingham
    and Ralph A. Griffiths
MEDIEVAL LITERATURE Elaine Treharne
MEDIEVAL PHILOSOPHY
    John Marenbon
MEMORY Jonathan K. Foster
METAPHYSICS Stephen Mumford
THE MEXICAN REVOLUTION
    Alan Knight
MICHAEL FARADAY
    Frank A. J. L. James
MICROBIOLOGY Nicholas P. Money
MICROECONOMICS Avinash Dixit
MICROSCOPY Terence Allen
THE MIDDLE AGES Miri Rubin
MILITARY JUSTICE Eugene R. Fidell
MINERALS David Vaughan
MODERN ART David Cottington
MODERN CHINA Rana Mitter
MODERN DRAMA
    Kirsten E. Shepherd-Barr
MODERN FRANCE Vanessa R. Schwartz

MODERN IRELAND Senia Pašeta
MODERN ITALY Anna Cento Bull
MODERN JAPAN
    Christopher Goto-Jones
MODERN LATIN AMERICAN
    LITERATURE
    Roberto González Echevarría
MODERN WAR Richard English
MODERNISM Christopher Butler
MOLECULAR BIOLOGY Aysha Divan
    and Janice A. Royds
MOLECULES Philip Ball
THE MONGOLS Morris Rossabi
MOONS David A. Rothery
MORMONISM Richard Lyman Bushman
MOUNTAINS Martin F. Price
MUHAMMAD Jonathan A. C. Brown
MULTICULTURALISM Ali Rattansi
MUSIC Nicholas Cook
MYTH Robert A. Segal
THE NAPOLEONIC WARS
    Mike Rapport
NATIONALISM Steven Grosby
NAVIGATION Jim Bennett
NELSON MANDELA Elleke Boehmer
NEOLIBERALISM Manfred Steger and
    Ravi Roy
NETWORKS Guido Caldarelli and
    Michele Catanzaro
THE NEW TESTAMENT
    Luke Timothy Johnson
THE NEW TESTAMENT AS
    LITERATURE Kyle Keefer
NEWTON Robert Iliffe
NIETZSCHE Michael Tanner
NINETEENTH-CENTURY BRITAIN
    Christopher Harvie and
    H. C. G. Matthew
THE NORMAN CONQUEST
    George Garnett
NORTH AMERICAN INDIANS
    Theda Perdue and Michael D. Green
NORTHERN IRELAND
    Marc Mulholland
NOTHING Frank Close
NUCLEAR PHYSICS Frank Close
NUCLEAR POWER Maxwell Irvine
NUCLEAR WEAPONS
    Joseph M. Siracusa
NUMBERS Peter M. Higgins
NUTRITION David A. Bender

OBJECTIVITY Stephen Gaukroger
THE OLD TESTAMENT
  Michael D. Coogan
THE ORCHESTRA D. Kern Holoman
ORGANIZATIONS Mary Jo Hatch
PANDEMICS Christian W. McMillen
PAGANISM Owen Davies
THE PALESTINIAN-ISRAELI
  CONFLICT Martin Bunton
PARTICLE PHYSICS Frank Close
PAUL E. P. Sanders
PEACE Oliver P. Richmond
PENTECOSTALISM William K. Kay
THE PERIODIC TABLE Eric R. Scerri
PHILOSOPHY Edward Craig
PHILOSOPHY IN THE ISLAMIC
  WORLD Peter Adamson
PHILOSOPHY OF LAW
  Raymond Wacks
PHILOSOPHY OF SCIENCE
  Samir Okasha
PHOTOGRAPHY Steve Edwards
PHYSICAL CHEMISTRY Peter Atkins
PILGRIMAGE Ian Reader
PLAGUE Paul Slack
PLANETS David A. Rothery
PLANTS Timothy Walker
PLATE TECTONICS Peter Molnar
PLATO Julia Annas
POLITICAL PHILOSOPHY David Miller
POLITICS Kenneth Minogue
POPULISM Cas Mudde and
  Cristóbal Rovira Kaltwasser
POSTCOLONIALISM Robert Young
POSTMODERNISM Christopher Butler
POSTSTRUCTURALISM Catherine Belsey
PREHISTORY Chris Gosden
PRESOCRATIC PHILOSOPHY
  Catherine Osborne
PRIVACY Raymond Wacks
PROBABILITY John Haigh
PROGRESSIVISM Walter Nugent
PROTESTANTISM Mark A. Noll
PSYCHIATRY Tom Burns
PSYCHOANALYSIS Daniel Pick
PSYCHOLOGY Gillian Butler and
  Freda McManus
PSYCHOTHERAPY Tom Burns and
  Eva Burns-Lundgren
PUBLIC ADMINISTRATION
  Stella Z. Theodoulou and Ravi K. Roy

PUBLIC HEALTH Virginia Berridge
PURITANISM Francis J. Bremer
THE QUAKERS Pink Dandelion
QUANTUM THEORY
  John Polkinghorne
RACISM Ali Rattansi
RADIOACTIVITY Claudio Tuniz
RASTAFARI Ennis B. Edmonds
THE REAGAN REVOLUTION Gil Troy
REALITY Jan Westerhoff
THE REFORMATION Peter Marshall
RELATIVITY Russell Stannard
RELIGION IN AMERICA Timothy Beal
THE RENAISSANCE Jerry Brotton
RENAISSANCE ART
  Geraldine A. Johnson
REVOLUTIONS Jack A. Goldstone
RHETORIC Richard Toye
RISK Baruch Fischhoff and John Kadvany
RITUAL Barry Stephenson
RIVERS Nick Middleton
ROBOTICS Alan Winfield
ROCKS Jan Zalasiewicz
ROMAN BRITAIN Peter Salway
THE ROMAN EMPIRE Christopher Kelly
THE ROMAN REPUBLIC
  David M. Gwynn
ROMANTICISM Michael Ferber
ROUSSEAU Robert Wokler
RUSSELL A. C. Grayling
RUSSIAN HISTORY Geoffrey Hosking
RUSSIAN LITERATURE Catriona Kelly
THE RUSSIAN REVOLUTION
  S. A. Smith
SAVANNAS Peter A. Furley
SCHIZOPHRENIA Chris Frith and
  Eve Johnstone
SCHOPENHAUER Christopher Janaway
SCIENCE AND RELIGION
  Thomas Dixon
SCIENCE FICTION David Seed
THE SCIENTIFIC REVOLUTION
  Lawrence M. Principe
SCOTLAND Rab Houston
SEXUALITY Véronique Mottier
SHAKESPEARE'S COMEDIES Bart van Es
SIKHISM Eleanor Nesbitt
THE SILK ROAD James A. Millward
SLANG Jonathon Green
SLEEP Steven W. Lockley and
  Russell G. Foster

SOCIAL AND CULTURAL
   ANTHROPOLOGY
   John Monaghan and Peter Just
SOCIAL PSYCHOLOGY Richard J. Crisp
SOCIAL WORK Sally Holland and
   Jonathan Scourfield
SOCIALISM Michael Newman
SOCIOLINGUISTICS John Edwards
SOCIOLOGY Steve Bruce
SOCRATES C. C. W. Taylor
SOUND Mike Goldsmith
THE SOVIET UNION Stephen Lovell
THE SPANISH CIVIL WAR
   Helen Graham
SPANISH LITERATURE Jo Labanyi
SPINOZA Roger Scruton
SPIRITUALITY Philip Sheldrake
SPORT Mike Cronin
STARS Andrew King
STATISTICS David J. Hand
STEM CELLS Jonathan Slack
STRUCTURAL ENGINEERING
   David Blockley
STUART BRITAIN John Morrill
SUPERCONDUCTIVITY
   Stephen Blundell
SYMMETRY Ian Stewart
TAXATION Stephen Smith
TEETH Peter S. Ungar
TELESCOPES Geoff Cottrell
TERRORISM Charles Townshend
THEATRE Marvin Carlson
THEOLOGY David F. Ford
THOMAS AQUINAS Fergus Kerr
THOUGHT Tim Bayne

TIBETAN BUDDHISM
   Matthew T. Kapstein
TOCQUEVILLE Harvey C. Mansfield
TRAGEDY Adrian Poole
TRANSLATION Matthew Reynolds
THE TROJAN WAR Eric H. Cline
TRUST Katherine Hawley
THE TUDORS John Guy
TWENTIETH-CENTURY BRITAIN
   Kenneth O. Morgan
THE UNITED NATIONS
   Jussi M. Hanhimäki
THE U.S. CONGRESS Donald A. Ritchie
THE U.S. SUPREME COURT
   Linda Greenhouse
UTOPIANISM Lyman Tower Sargent
THE VIKINGS Julian Richards
VIRUSES Dorothy H. Crawford
VOLTAIRE Nicholas Cronk
WAR AND TECHNOLOGY Alex Roland
WATER John Finney
WEATHER Storm Dunlop
THE WELFARE STATE David Garland
WILLIAM SHAKESPEARE
   Stanley Wells
WITCHCRAFT Malcolm Gaskill
WITTGENSTEIN A. C. Grayling
WORK Stephen Fineman
WORLD MUSIC Philip Bohlman
THE WORLD TRADE
   ORGANIZATION Amrita Narlikar
WORLD WAR II Gerhard L. Weinberg
WRITING AND SCRIPT
   Andrew Robinson
ZIONISM Michael Stanislawski

Available soon:

INFINITY Ian Stewart
ORGANIC CHEMISTRY
   Graham Patrick
THE ATMOSPHERE Paul I. Palmer

SHAKESPEARE'S TRAGEDIES
   Stanley Wells
CLINICAL PSYCHOLOGY
   Susan Llewelyn and
   Katie Aafjes-van Doorn

For more information visit our website

www.oup.com/vsi/

Russell G. Foster and Leon Kreitzman

CIRCADIAN
RHYTHMS

A Very Short Introduction

OXFORD
UNIVERSITY PRESS

OXFORD
UNIVERSITY PRESS

Great Clarendon Street, Oxford, OX2 6DP,
United Kingdom

Oxford University Press is a department of the University of Oxford.
It furthers the University's objective of excellence in research, scholarship,
and education by publishing worldwide. Oxford is a registered trade mark of
Oxford University Press in the UK and in certain other countries

First edition published in 2017

Impression: 8

Published in the United States of America by Oxford University Press
198 Madison Avenue, New York, NY 10016, United States of America

British Library Cataloguing in Publication Data
Data available

Library of Congress Control Number: 2016955251

ISBN 978-0-19-871768-3

Printed and bound by CPI Group (UK) Ltd, Croydon, CR0 4YY

Contents

Acknowledgements xv

Foreword xvii

List of illustrations xix

1 Circadian rhythms: A 24-hour phenomenon 1

2 Time of day matters 13

3 When timing goes wrong 24

4 Shedding light on the clock 45

5 The tick-tock of the molecular clock 62

6 Sleep: The most obvious 24-hour rhythm 81

7 Circadian rhythms and metabolism 91

8 Seasons of life 106

9 Evolution and another look at the clock 122

Further reading 131

Index 137

Acknowledgements

Many colleagues have provided advice and feedback on our manuscript. We would particularly like to thank Charalambos ('Bambos') Kyriacou, Professor of Behavioural Genetics at the University of Leicester, for his input on Chapter 5. In addition, we would like to thank Dr Aarti Jagannath and Dr Sridhar Vasudevan for helpful discussions and feedback on the entire manuscript.

Foreword

Circadian rhythms are found in nearly every living thing on earth. They help organisms time their daily and seasonal activities so that they are synchronized to the external world and the predictable changes in the environment. These biological clocks provide a cross-cutting theme in biology and they are incredibly important. They influence everything, from the way growing sunflowers track the sun from east to west, to the migration timing of monarch butterflies, to the morning peaks in cardiac arrest in humans.

Despite the diversity of life on our planet, there are many similarities in the way in which circadian rhythms are generated and synchronized to the solar cycle. There is a molecular feedback loop—the transcription–translation feedback loop (TTFL)—that underpins all these processes, and our understanding of this molecular clockwork provide the best example to date of how genes and their protein products interact to generate complex behaviour.

Circadian rhythms are found in bacteria, algae, fungi, plants, and animals, but we have had to concentrate our discussion on mammals. Although rats and mice are familiar, the terminology used to describe the circadian rhythms in these species may be alien to readers, and it is this terminology that makes some of the diagrams seem daunting, but the concepts are, we hope, much easier to follow.

Science is about perseverance. Years of work underlie most scientific discoveries. Explaining these discoveries in a way that can be understood is not always easy. We have tried to keep the general reader in mind but in places perseverance on the part of the reader may be required. In the end we were guided by one of our reviewers, who said: 'If you want to understand calculus you have to show the equations.' We hope you will find the effort is worth it, as these rhythms are one of the signatures of life and an understanding of these biological processes tells us much about ourselves and the world in which we live.

List of illustrations

1 The basic terminology used to describe biological oscillations, such as circadian rhythms **4**

2 Leaf movements recorded using a kymograph drum **8**

3 The building blocks of circadian systems **10**

4 An illustration of the average 24-hour variation in different parameters of adult human physiology and behaviour **15**

5 Examples of circadian changes in adult human physiology and behaviour that persist in the laboratory under constant conditions **16**

6 There is a near-normal or Gaussian distribution of chronotypes across the population, but with a slight over-representation of late types **18**

Based on figure 1 in Roenneberg and Merrow, *Circadian Entrainment of Neurospora crassa, Cold Spring Harbor Symposia on Quantitative Biology LXXII* (Cold Spring Harbor Laboratory Press, 2007), by permission. Bird images © iStock.com/Elena Belous

7 Normal versus abnormal patterns in sleep timing **34**

8 The possible relationships between psychiatric illness and sleep and circadian rhythm disruption (SCRD) **39**

9 The phase response curve (PRC) for a nocturnal animal such as a mouse **48**

10 Rods and cones convey visual information to the retinal ganglion cells via the second-order neurons of the inner retina—the bipolar and amacrine neurons **51**

11 Diagram of the rat brain showing the dedicated retino-hypothalamic tract (RHT) projection from the pRGCs of the eye to the suprachiasmatic nucleus (SCN) which contains the master circadian pacemaker of mammals, and the paired SCN located either side of the third ventricle (III) which sit on top of the optic chiasm (OC) **54**

12 An early version of *Drosophila* transcription–translation feedback loop (TTFL) **68**

13 Outline of the current working model of the circadian molecular clockwork of the fruit fly *Drosophila* **70**

14 The current working model of the circadian molecular clockwork of the mouse **74**

15 The light regulation of the molecular clockwork in mammals **76**

16 The two-process model of sleep regulation for a diurnal mammal **85**

17 Sleep/wake states arising from mutually excitatory and inhibitory circuits **87**

18 The key interactions in glucose metabolism and the responses involved in low blood glucose (hypoglycaemia) and high blood glucose (hyperglycaemia) **93**

19 The links between the circadian system and hormonal and metabolic rhythms in nocturnal and diurnal mammals **96**

20 The transcriptional control of metabolic pathways by molecular clocks **103**

21 The Earth's axis **107**

22 Bünning's hypothesis, or the 'external coincidence model' of photoperiodic regulation **110**

23 A model of how photoperiod regulates flowering in *Arabidopsis* **112**

24 Summary of the key events involved in photoperiodic time measurement in birds and mammals stimulated by the increasing daylengths of spring (long-day breeders) **117**

25 A circadian oscillation in cyanobacteria **127**

Chapter 1
Circadian rhythms: A 24-hour phenomenon

There have been over a trillion dawns and dusks since life began some 3.8 billion years ago. During that time the earth's daily rotation has slowed to a shade less than 24 hours—or 23 hours 56 minutes and 4 seconds to be precise. This predictable daily solar cycle results in regular and profound changes in environmental light, temperature, and food availability as day follows night. Almost all life on earth, including humans, employs an internal biological timer to anticipate these daily changes. The possession of some form of clock permits organisms to optimize physiology and behaviour in advance of the varied demands of the day/night cycle. Organisms effectively 'know' the time of day. Such internally generated daily rhythms are called 'circadian rhythms' from the Latin *circa* (about) and *dies* (day).

Organisms that use circadian rhythms to anticipate the rotation of the earth are thought to have a major advantage over both their competitors and predators. For example, it takes about 20–30 minutes for the eyes of fish living among coral reefs to switch vision from the night to daytime state. A fish whose eyes are prepared in advance for the coming dawn can exploit the new environment immediately. The alternative would be to wait for the visual system to adapt and miss out on valuable activity time, or emerge into a world where it would be more difficult to avoid predators or catch prey until the eyes have adapted. Efficient use of time to maximize survival almost certainly provides a large selective advantage, and consequently all organisms seem to be led by such anticipation.

A circadian clock also stops everything happening within an organism at the same time, ensuring that biological processes occur in the appropriate sequence or 'temporal framework'. For cells to function properly they need the right materials in the right place at the right time. Thousands of genes have to be switched on and off in order and in harmony. Proteins, enzymes, fats, carbohydrates, hormones, nucleic acids, and other compounds have to be absorbed, broken down, metabolized, and produced in a precise time window. Energy has to be obtained and then partitioned across the cellular economy and allocated to growth, reproduction, metabolism, locomotion, and cellular repair. All of these processes, and many others, take energy and all have to be timed to best effect by the millisecond, second, minute, day, and time of year. Without this internal temporal compartmentalization and its synchronization to the external environment our biology would be in chaos.

The imposition of an internal temporal framework, combined with the anticipation of environmental change, makes the possession of a circadian timing system an essential part of an organism's biology. The mechanisms underlying circadian rhythms involve circadian oscillations in gene expression, protein modifications, and ultimately behaviour. These oscillations are controlled by the signals generated by the levels of the core clock genes. However, to be biologically useful, these rhythms must be synchronized or entrained to the external environment, predominantly by the patterns of light produced by the earth's rotation, but also by other rhythmic changes within the environment such as temperature, food availability, rainfall, and even predation. These entraining signals, or time-givers, are known as *zeitgebers*. The key point is that circadian rhythms are not driven by an external cycle but are generated internally, and *then* entrained so that they are synchronized to the external cycle.

Circadian rhythms are embedded within the genomes of just about every plant, animal, fungus, algae, and even cyanobacteria (a phylum of photosynthetic eubacteria). Our own daily patterns of sleeping and waking, eating and drinking, depend not just on an alarm clock nor how much exercise we have done, but fundamentally on what our internal biological clocks are instructing us to do. When left without time cues, such as deep underground or during an Arctic winter or in an experimental isolation chamber, our endogenous clocks still tick and attempt to drive us.

Human physiology is organized around the daily cycle of activity and sleep. In the active phase, when energy expenditure is higher and food and water are consumed, organs need to be prepared for the intake, processing, and uptake of nutrients. The activity of organs such as the stomach, liver, small intestine, pancreas, and blood flow to these organs needs internal synchronization—which a clock can provide. Sleep may be the suspension of most physical activity, but during sleep many essential activities occur including: cellular repair, the removal of toxins, and in the brain, memory consolidation and information processing. Disrupting this pattern, as happens with jet lag or shift work, leads to internal desynchrony of the circadian system and our ability to do the right thing at the right time is greatly impaired.

Despite the incredible diversity of biological clocks and rhythms, they can all be depicted as a single hand sweeping around the face of a clock once, but not exactly, every 24 hours as illustrated in Figure 1. The sinusoidal wave is a graphical depiction of the rhythm and represents the rise and fall in the cellular levels of molecules under circadian control which are as diverse as mRNAs, enzymes, proteins, and neurotransmitters.

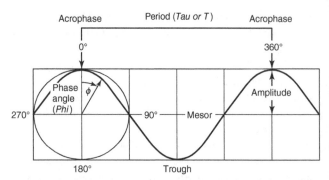

1. The basic terminology used to describe biological oscillations, such as circadian rhythms. The period for most circadian rhythms under constant conditions of light or temperature is not exactly 24 hours but either a little longer or shorter.

The defining period of the oscillation (also called *Tau* or *T*) is the time taken for one revolution (360°) between successive reference points (phase points) on the cycle. Peaks on the cycle are called acrophases. The extent of the increase (or decrease) in the biological parameter is called the amplitude compared to the mesor (the average value around which the variable oscillates). Some also define the amplitude as the peak to trough variation. The phase angle (*Phi* or Φ) is less intuitive, but imagine the hand of a clock on the face of the dial. If the hand is extended to the curve on the clock face there is an angle between it and the starting point. This is the phase angle or the angular displacement between a point on the oscillation and a reference angle.

In constant conditions, such as constant light or constant darkness, and with no exposure to entraining zeitgebers, circadian rhythms *freerun*. Depending on the species, the period can vary between 22 and 25 hours, although there is considerable variation both between species and between individuals of the same species. Most humans have slightly long body clocks and freerun with a period of about 24 hours and 10 minutes.

4

The ability to show sustained near 24-hour rhythmic activity under constant conditions for several cycles is considered to be one of the key defining features of a true circadian rhythm, as distinct from a 24-hour pattern of activity that is simply driven by changes in the environment. That circadian rhythms continue for many cycles under constant conditions is surprising, as most life will never or rarely have experienced constant conditions of light or dark, with the notable exception of animals that live in the Arctic, Antarctic, underground, in the deep sea, or within the dark interior of a hive or termite mound.

Honeybee behaviour is a classic example of circadian timing. They are acutely attuned to the way in which their environment changes as they go about collecting nectar and pollen from flowers. Plants have to attract pollinators—they cannot go to them! Over a 24-hour cycle, plants maximize the production of nectar, pollen, and scents so that they are available during a particular time of day. All the information about a flower is time-linked, and flowers of a given species all produce nectar at about the same time each day. This concentrates foraging to a particular species over a narrow time window which increases the probability of cross-pollination.

Bees and flowering plants have co-evolved the synchronization of the timing of a critical daily activity. Their internal clocks enable them to anticipate the future and prepare for it. In effect the bees have a daily appointments book for flower-visiting and they can 'remember' as many as nine appointments a day. The bees turn up at a particular time, receive their food reward, and in turn help the plants cross-pollinate. Both bees and plants share a common internal representation of the solar day and they can 'tell' the time and synchronize their internal 'watches'.

Social insects such as honeybees, ants, wasps, and termites live in colonies consisting of many thousands of individuals. Workers in these complex societies specialize in different tasks that need tight temporal coupling. This coordination of their activities is important for an efficient colony and, as a result, the fitness of genetically related individuals within the colony. One example of the complex temporal relationships is the specialized nectar reception among honeybees. Forager bees transfer the nectar they collect during the day to a specialized group of 'nectar receivers'. The nectar receiver bees accept the nectar load, find empty cells, and store the nectar that will later be processed into honey. The nectar receivers typically stay inside the dark and temperature-stable hive. They lack light and temperature time cues but still 'know' the correct time because the foragers, who are outside and are exposed to time cues, entrain the interior workers by some form of social interaction. In this case the foragers act as zeitgebers for the receivers.

Although organisms use their circadian timing systems to anticipate the predictable 24-hour world, the system needs to be flexible. Newly emerged bees typically have no circadian rhythms in locomotion or metabolic activity. Circadian rhythms appear only later in their short lives. Honeybees care for ('nurse') the brood during the second and third weeks of their adult life, and because this is a continual 24/7 activity they show no overt circadian rhythms. When they move into the foraging role outside the nest they show strongly entrained circadian rhythms and consolidated periods of rest during the night. This flexibility or 'plasticity' in regard to their rhythms enables the bees to act out their roles to best effect in a dynamic temporal environment.

Some fish species can show either diurnal or nocturnal behaviour, and shift from one to the other depending on the season and developmental stage. Such circadian plasticity is of particular importance to animals living in the polar regions. Species such as the Arctic reindeer live in regions where photoperiodic (day length) information (Chapter 8) is much reduced or even absent for considerable periods. These animals are exposed to continuous daylight in the summer months and darkness in the winter.

During these periods the clock function that drives circadian rhythms is much reduced if not wholly absent. Switching off the clock probably helps them maximize food intake. During the summer, sustained feeding off the abundance of vegetation will allow the development of food reserves in preparation for the severe winter conditions, while in winter reindeer will be able to graze whenever the harsh weather conditions permit.

We have been aware of circadian rhythms for thousands of years without realizing what they were, how they worked, or how crucial they are for our health and well-being. In the 4th century BC, Androsthenes, one of Alexander the Great's sea captains, described the daily leaf movements of the tamarind tree (*Tamarindus indicus*), as they curled and uncurled. The medical pioneers Hippocrates and Galen noted the periodic 24-hour rhythmicity associated with fever. But although many observations were made throughout the centuries, indicating that plants and animals carry out their activities in regularly timed 24-hour cycles, these were no more than interesting facts of nature. What began to change this view was a remarkably simple biological experiment published in 1729 by a French astronomer, Jean Jacques d'Ortous de Mairan.

De Mairan knew that a mimosa plant (probably *Mimosa pudica*) was a heliotrope, turning towards the sun in daylight; but as well as being sensitive to the sun's position it was sensitive in another way—its leaves would droop at dusk and rise during the day. De Mairan put a mimosa plant in a cupboard to see what happened when it was kept in the dark. He peeked in at various times and observed that its leaves still opened and closed rhythmically—it was as though it had its own representation of day and night. The plant's leaves still drooped when the plant *considered* it to be night (subjective night) and rose up during the plant's subjective day. He had unwittingly shown that the plant had an endogenous rhythm which persisted under constant conditions; in effect it had its own internal biological clock. It took another 230 years to come up with the term to label this rhythm and nearly as long for the endogenous nature of such rhythms to be finally accepted.

The American physiologist Curt Richter in the 1920s was the first to demonstrate that the 24-hour 'drive' for sleep/wake behaviour in rodents is endogenous. He was able to show, working mainly with wild-caught rats using home-built running wheels, that the animals had their own rhythms independent of the direct influences of the environment. In 1930s Germany, Erwin Bünning laboriously hooked up the leaves of *Phaseolus coccineus* (the runner bean) with thread to a kymograph (Figure 2) and plotted the movements of the leaves in light/dark cycles and also in constant light or dark conditions. As illustrated in Figure 2, the pressure and movement of a lever pressing on to a 24-hour rotating drum (covered with waxed or sooty paper) leaves a trace of the rising and lowering of the leaf. During the first day 0 > 1, the plant is exposed to a light/dark cycle with leaves raised during the light. For the following days 1 > 6, the plant is kept under

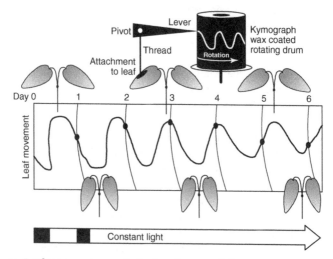

2. **Leaf movements recorded using a kymograph drum.**

constant light. The rhythm in leaf movement continues under constant conditions but the period lengthens, with leaves rising and falling later with respect to the astronomical day (0–6). Note the leaf position indicated by the black dot at the end of each day. By the mid-1930s Bünning had shown that in plants at least the 'biological clock', a term he coined, was endogenous.

It is worth emphasizing that the concept of an internal clock, as developed by Richter and Bünning, has been enormously powerful in furthering our understanding of biological processes in general, providing a link between our physiological understanding of homeostatic mechanisms, which try to maintain a constant internal environment despite unpredictable fluctuations in the external environment (e.g. after a temperature change as clouds pass overhead or recovering from running away from a predator), versus the circadian system which enables organisms to anticipate periodic changes in the external environment. The circadian system provides a predictive 24-hour baseline in physiological parameters, which is then either defended or temporarily overridden by homeostatic mechanisms that accommodate an acute environmental challenge.

As illustrated in Figure 3, the building blocks of any circadian system consist of a 24-hour central pacemaker or 'clock' which sits at the heart of the circadian system. Zeitgebers and the entrainment pathway synchronize the internal day to the astronomical day, usually via the light/dark cycle, and multiple output rhythms in physiology and behaviour allow appropriately timed activity. The multitude of clocks within a multicellular organism can all potentially tick with a different phase angle (Figure 1), but usually they are synchronized to each other and by a central pacemaker which is in turn entrained to the external world via appropriate zeitgebers.

The simple model in Figure 3 has been a key metaphor in biological clock research and is still a very useful conceptual tool

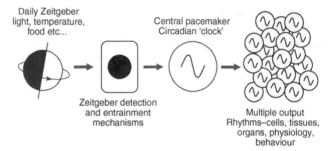

Daily Zeitgeber
light, temperature,
food etc...

Central pacemaker
Circadian 'clock'

Zeitgeber detection
and entrainment
mechanisms

Multiple output
Rhythms–cells, tissues,
organs, physiology,
behaviour

3. The building blocks of circadian systems: a self-sustained 24-hour rhythm generator (oscillator), a setting (entraining) mechanism that links the endogenous oscillation to the astronomical oscillations, such as dawn and dusk, and a means of making the output rhythms biologically useful.

to organize both thought and exposition. Circadian rhythms are encoded in our genes, endogenously generated, and self-sustaining with a period of about 24 hours that persists under constant environmental conditions. But the circadian system requires an additional key feature. The circadian period must remain relatively constant over a range of ambient temperatures. The clock has to be 'temperature compensated', and this represents yet another extraordinary feature of circadian clocks. Most biological reactions vary greatly with temperature and show a Q10 temperature coefficient of about 2 (Q10 = 2). This means that the biological process or reaction rate doubles as a consequence of increasing the temperature by 10°C up to a maximum temperature at which the biological reaction stops. For example, the velocity and power of muscle contraction declines with decreasing temperatures and increases with rising temperatures, such that a 10°C temperature increase doubles muscle performance. By contrast, circadian rhythms exhibit a Q10 close to 1 (Q10 = 1).

Clocks without temperature compensation are useless. When John Harrison built the mechanical clocks that won him the

Longitude Prize he used metals with different coefficients of expansion to compensate for temperature change. His clocks did not speed up or slow down when carried on a ship anchored in London or the Caribbean. Although we know that circadian clocks show temperature compensation, and that this phenomenon is a conserved feature across all circadian rhythms, we have little idea how this is achieved.

The systematic study of circadian rhythms only really started in the 1950s, and the pioneering studies of Colin Pittendrigh bought coherence to this emerging new discipline. Pittendrigh, a British-born biologist who worked for most of his life in the USA, studied circadian rhythms across physiology and behaviour from the hatching (eclosion) of fruit flies (*Drosophila*) to mouse activity patterns, and encouraged others to work on everything from unicellular organisms to humans. From this mass of emerging data, Pittendrigh had key insights and defined the essential properties of circadian rhythms across all life. Namely that: all circadian rhythms are endogenous and show near 24-hour rhythms in a biological process (biochemistry, physiology, or behaviour); they persist under constant conditions for several cycles; they are entrained to the astronomical day via synchronizing zeitgebers; and they show temperature compensation such that the period of the oscillation does not alter appreciably with changes in environmental temperature. Much of the research since the 1950s has been the translation of these formalisms into biological structures and processes, addressing such questions as: What is the clock and where is it located within the intracellular processes of the cell? How can a set of biochemical reactions produce a regular self-sustaining rhythm that persists under constant conditions and has a period of about 24 hours? How is this internal oscillation synchronized by zeitgebers such as light to the astronomical day? Why is the clock not altered by temperature, speeding up when the environment gets hotter and slowing down in the cold? How is the information of the near 24-hour rhythm communicated to the rest of the organism?

The ubiquity of circadian clocks and rhythms speaks to their importance. They play a key role in a vast array of physiological processes, including sleep/wake cycles, sexual behaviour and reproduction, thermoregulation, and metabolic control such as energy intake/expenditure, glucose metabolism, lipid metabolism, and food and water intake. Such innate timing mechanisms have been a feature of life on this planet since the first cyanobacteria and their production of oxygen began to shape the future of the planet billions of years ago. Circadian rhythms not only have a large impact upon us, but when they are disrupted our overall health and well-being can be severely affected.

Chapter 2
Time of day matters

Serious interest in how performance changes across the day goes back to the end of the 19th century. In 1892, Fletcher Bascom Dresslar, a research student at Clarke University in Massachusetts, published a paper on 'Some Influences which Affect the Rapidity of Voluntary Movements'. Dresslar tapped 300 times on a Morse key used by telegraph operators, timing himself in the process. He started tapping at 08.00 and repeated the procedure every two hours until 18.00 every day for six weeks. His tapping got faster from 08.00 to 12.00, slowed after lunch, but then picked up speed again and was quicker at 18.00 than at 08.00.

There have been hundreds of studies showing that a broad range of activities, both physical and cognitive, vary across the 24-hour day: tooth pain is lowest in the morning; proofreading is best performed in the evening; labour pains usually begin at night and most natural births occur in the early morning hours. The accuracy of short and long badminton serves is higher in the afternoon than in the morning and evening. Accuracy of first serves in tennis is better in the morning and afternoon than in the evening, although speed is higher in the evening than in the morning. Swimming velocity over 50 metres is higher in the evening than in the morning and afternoon. Studies of soccer skills have reported better performances in the evening for the wall-volley test and dribbling speed. The majority of studies report that performance increases from morning to afternoon or evening. Sports involving technical skills (e.g. badminton, tennis, soccer) seem to peak somewhat earlier in the day than those of muscle strength and anaerobic performance, which seem to peak in the early evening.

Examples of physiological and cognitive change in adults across the 24-hour day are illustrated in Figure 4. It is likely that most of these changes are driven in large part by the circadian system, but a formal demonstration that these rhythms persist under constant conditions has only been made for a small number of these changes. For such experiments individuals would need to be maintained in isolation and without 24-hour time cues for several days.

In contrast to Figure 4, Figure 5 illustrates several examples of 24-hour physiology and behaviour that have been shown to persist under constant conditions. Day versus night variations in blood pressure and heart rate are among the best-known circadian rhythms of physiology. In humans, there is a 24-hour variation in blood pressure with a sharp rise before awakening (Figure 5). Many cardiovascular events, such as sudden cardiac death, myocardial infarction, and stroke, display diurnal variations with an increased incidence between 06.00 and 12.00 in the morning. Both atrial and ventricular arrhythmias appear to exhibit circadian patterning as well, with a higher frequency during the day than at night.

Myocardial infarction (MI) is two to three times more frequent in the morning than at night. In the early morning, the increased systolic blood pressure and heart rate results in an increased energy and oxygen demand by the heart, while the vascular tone of the coronary artery rises in the morning, resulting in a decreased coronary blood flow and oxygen supply. This mismatch between supply and demand underpins the high frequency of onset of MI. Plaque blockages are more likely to occur in the

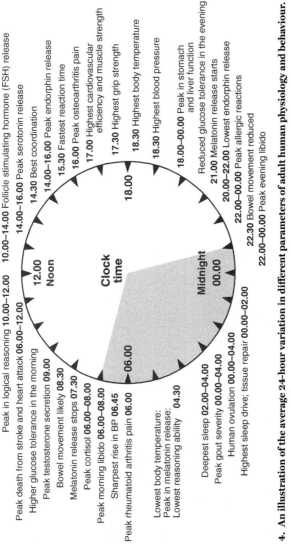

4. **An illustration of the average 24-hour variation in different parameters of adult human physiology and behaviour.**

15

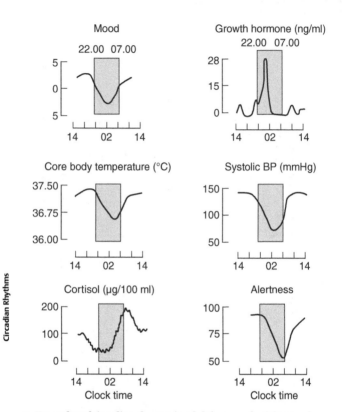

5. Examples of circadian changes in adult human physiology and behaviour that persist in the laboratory under constant conditions. The grey vertical bar indicates the approximate sleep time between 22.00 and 07.00.

morning as platelet surface activation markers have a circadian pattern producing a peak of thrombus formation and platelet aggregation. The resulting hypercoagulability partially underlies the morning onset of MI.

Several studies on a variety of animal models suggest that circadian rhythms can be very important in information processing. In one study the fruit fly (*Drosophila melanogaster*) learned to associate an electric shock with a particular odour. The flies' learning was measured in a T-maze as the avoidance of the odour associated with the electric shock compared to a 'harmless' odour. Memory formation in the flies was regulated in a circadian manner with peak performance in short-term memory during the early evening. This rhythm in memory was abolished in mutant flies lacking a circadian clock. Work on sea slugs, fish, and rodents has produced similar findings. Simply put, an organism is set up to function differently at different times of the day. So the diurnal sea slug *Aplysia californica* exhibits greater memory formation for associative learning during the day, while the nocturnal *Aplysia fasciata* is more capable of memory formation at night.

Perhaps the most obvious and profound of our 24-hour rhythms is the cycle of activity and sleep (Chapter 6). Till Roenneberg, based in Munich, has investigated the sleep timing of individuals from around the world. He and his colleagues have utilized an internet-based questionnaire (the Munich ChronoType Questionnaire, MCTQ) that assessed preferred sleep and wake times on work days and on free days, and have collected data from well over 200,000 respondents. The MCTQ is generally considered to give the best estimate of morningness and eveningness chronotypes in normal adults because it defines sleep on free days without the influence of alarm-clock-driven wake times. Mid-sleep-time is calculated as the mid-point between reported bed and wake times.

Roenneberg's studies have developed the notion of an individual chronotype. Originally chronotype was used to categorize individuals as either larks, owls, or neither. For example, using the original classification schemes about 10–15 per cent of us are morning people, or larks, starting our days early. Another 10–15 per cent or so are owls, who find rising early very difficult and go to bed late. Roenneberg has shown that there is a spectrum of chronotypes ranging from extreme early types, who sleep (without

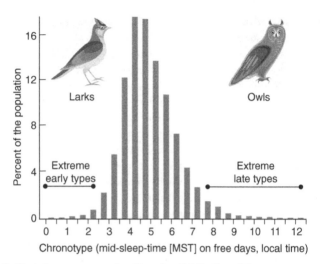

6. There is a near-normal or Gaussian distribution of chronotypes across the population, but with a slight over-representation of late types.

social obligations) from 20.00 to 04.00, to extreme late types, who sleep from 04.00 to 12.00 noon (Figure 6).

An important finding from the MCTQ showed that chronotype is also influenced by the natural light/dark cycle within a particular time zone. In the same time zone (e.g. Central European Time, used from eastern Poland to western Spain) chronotype moves from more early types in the east to more late types in the west. Even across Germany, despite the population experiencing similar social times (work times, evening news programmes, etc.), the internal timing system of the inhabitants is still adjusted to sunrise. The chronotype becomes later by four minutes for every longitude degree from east to west, the same time that the sun takes to cross one longitude.

The MCTQ data have shown that many of us live on two timetables, one enforced by the weekday alarm clock on work days, and the other more aligned to our internal circadian time, involving 'sleeping in' at the weekend. To align sleep and wake times with social obligations, 80 per cent of the population uses an alarm clock on workdays, and a growing number of people use sleep medication at night and stimulants to drive wakefulness during the day.

The difference between alarm-clock-driven wake times and natural wake times on free days has been termed by Roenneberg 'social jet lag'. An extreme late type, who is forced to get up and go to work at 07.00, will experience massive social jet lag. The same would be true for an extreme morning type who, under peer pressure, is forced to stay up socializing on a Friday until the early hours of the morning.

Only 13 per cent of the population represented in the MCTQ database is free of social jet lag, 69 per cent experience at least 1 hour, and a third suffers from 2 hours or more. The discrepancy between biological (circadian) time and social time is a major reason for sleep deprivation. About 5 per cent of the population in the database sleep at least 20 per cent fewer hours than their biological need. An additional 35 per cent sleep up to 10 per cent less on workdays, missing half a night's sleep every week. Only 25 per cent of the population gets the same amount of sleep on workdays and free days.

Determining individual chronotypes can be used to mitigate social jet lag and some of the disruptive effects of shift working (Chapter 3). Evening-type individuals report better tolerance for night work (e.g. better work performance and higher job satisfaction) than morning-type individuals. But social jet lag is associated with increased health risks. Independent of social background or region, the number of smokers in the population increases with the amount of social jet lag, and the consumption of alcohol and caffeine increases, along with increased levels of depression.

Being forced to live against one's circadian clock has metabolic consequences. Social jet lag is associated with an increased body mass index: with every hour of social jet lag the probability of being overweight or obese increases by 30 per cent. By contrast low levels of social jet lag, as in morning types, are correlated with greater levels of cheerfulness and alertness. One suggested reason why early risers may be happier is because their biological clocks are more in line with societal expectations about when someone should wake up and go to sleep!

Although an individual's chronotype is profoundly influenced by genetics, and where you live in a time zone, chronotype is not fixed through development. The clock gets later (has a later phase) from late childhood through adolescence, reaching peak lateness in women at 19.5 and in men at 21, meaning that young adults tend to want to stay in bed in the morning. With age, individuals move to an earlier phase so that by the time we are in our late fifties and early sixties our sleep timing resembles that of late childhood.

Typical 'optimal' times of day for physical or cognitive activity are gathered routinely from population studies (as shown in Figure 4). However, there is considerable individual variation. Peak performance will depend upon age, chronotype, time zone, and for behavioural tasks how many hours the participant has been awake when conducting the task, and even the nature of the task itself. As a general rule, the circadian modulation of cognitive functioning results in an improved performance over the day for younger adults, while in older subjects it deteriorates. A classic study by Lynn Hasher and colleagues at the University of Toronto compared cognitive performance at two times, mid-morning and mid-afternoon, in teenagers and adults. Test scores in teenagers increased by 10 per cent from mid-morning to mid-afternoon, while the adults declined by 7 per cent. These findings highlight an important dilemma. Senior teachers in their fifties will generally be at their best first thing in the morning, but their students will invariably be ill-prepared by their circadian system to learn. This finding may explain why there is a widespread belief that demanding subjects such as mathematics and science should be taught early in the school day while other, possibly less demanding topics such as physical education, art, and music, should be taught later in the day. The senior teachers and not the students determine the timetable and the tacit assumption for well over a century has been that students are most alert in the morning and the most important and intellectually demanding subjects should naturally be taught during this time. This assumption is wrong.

On average the circadian rhythms of an individual in their late teens will be delayed by around two hours compared with an individual in their fifties. As a result the average teenager experiences considerable social jet lag, and asking a teenager to get up at 07.00 in the morning is the equivalent of asking a 50-year-old to get up at 05.00 in the morning. Teenagers are biologically predisposed to get up late and go to bed late, but this predisposition has been exaggerated in recent years because societal/parental attitudes towards 'bedtime' have become more relaxed and bedrooms have been transformed from places of sleep to places of entertainment, packed full of electronic devices and 24/7 access to the internet which act to delay the onset of sleep further. Bedtimes are pushed back later, but the alarm on a school day still goes off at the same time. It has been estimated by Mary Carskadon at Brown University in the USA that for full cognitive performance teenagers need about 9 hours sleep each night, yet many get less than 6.5 hours.

In the USA, such observations prompted several schools to implement a delayed start to classes along with education relating to the importance and regulation of sleep (sleep hygiene/ education). After the later start, academic performance and attendance increased, while depression and self-harm declined. Such findings are consistent with a small study in the UK where start times were shifted from 08:50 to 10:00. This led to an increase in the percentage of pupils crossing an attainment threshold from 35 per cent to 53 per cent, and in those children classified as socially disadvantaged the increase was from 12 per cent to 42 per cent. This small project is now being followed up by a major study involving hundreds of students across the UK to assess the impact of both a later start time and sleep education on academic performance and student well-being. The results should finally provide an answer to the old question of what time the school day should start.

Chronotype also seems to play a key role in athletic performance. It had been accepted that performance peaks in the evening. However, as Roland Brandstaetter at the University of Birmingham in the UK has pointed out, previous research had measured and then averaged athletic performance together at different times, ignoring individual chronotypes. When chronotype is assessed, morning types were shown to have their peak performances at midday, the intermediate group did best in the afternoon, and the evening types did best in the evening. Everyone did the worst at 07.00 a.m. In the most extreme cases, people who naturally go to bed late and wake up late were as much as 26 per cent slower when they ran in the morning compared to the evening, even while trying as hard as they could. When all the data from the athletes were averaged, ignoring chronotype, the group performed best in the evening. It was only when the athletes were assessed on the basis of their chronotype that the striking differences emerged. So along with diet and motivation coaching, circadian coaching for athletes is likely to be integrated into future training regimes.

A critical area where time of day matters to the individual is the optimum time to take medication, a branch of medicine that has been termed 'chronotherapy'. Statins are a family of cholesterol-lowering drugs which inhibit HMGCR reductase (3-hydroxy-3-methyl-glutaryl-CoA reductase), the rate-limiting enzyme in endogenous cholesterol biosynthesis. Statins reduce the production of low-density lipoprotein (LDL-C). HMGCR is under circadian control and is highest at night. Hence those statins with a short half-life, such as simvastatin and lovastatin, are most effective when taken before bedtime. In another clinical domain entirely, recent studies have shown that anti-flu vaccinations given in the morning provoke a stronger immune response than those given in the afternoon.

The idea of using chronotherapy to improve the efficacy of anti-cancer drugs has been around for the best part of 30 years. Francis Lévi, now based at Warwick University in the UK, pioneered early work which showed that by varying the timing of medication to the individual, higher doses of anti-cancer drugs could be administered. One of the challenges in cancer therapy is the side effect of treatment. Tumours are generally fast-dividing cells and the medication attempts to kill rapidly dividing cells. But hair follicles and the endothelial lining of the gut are also fast-dividing, hence the side effects of hair loss and nausea. By working out the circadian timing of the follicle and endothelial division, therapeutic doses can be given when they should have maximal effect on the tumour cells and minimal effect on other cells. In experimental models more than thirty anti-cancer drugs have been found to vary in toxicity and efficacy by as much as 50 per cent as a function of time of administration.

Although Lévi and others have shown the advantages to treating individual patients by different timing regimes, few hospitals have taken it up. One reason is that the best time to apply many of these treatments is late in the day or during the night, precisely when most hospitals lack the infrastructure and personnel to deliver such treatments.

There is still a long way to go, but we are on the verge of very different medical care for cancer patients, where personalized treatments will be tailored to the most effective time of the day based upon the circadian chronotype of the patient and the timing of cell division within the tumour—the chronotype of the cancer.

Chapter 3
When timing goes wrong

While time of day, interacting with an individual's chronotype, can have an important impact upon performance and health, severe disruption of the circadian system adds another level of complexity and severity. Flying across multiple time zones and shift work has significant economic benefits, but the costs in terms of ill health are only now becoming clear. Sleep and circadian rhythm disruption (SCRD) is almost always associated with poor health.

Jet lag

During the Cold War of the 1950s, American Secretary of State John Foster Dulles made a long-haul flight to Cairo on a military jet to discuss US funding of the Aswan Dam. Dulles found it hard to concentrate in the meetings. When he arrived back in Washington after another long flight he learned that the Egyptians had just bought a substantial amount of Russian arms. Without due reflection, Dulles immediately cancelled his agreement with Colonel Nasser to bankroll the project. The Aswan Dam was built by Soviet money and engineers and this gave the Soviet Union their first foothold in Africa. At the time, Dulles put his failure to broker a lasting deal with Egypt down to travel fatigue. He can be forgiven for not being aware of the damaging effects of disrupting circadian rhythms by crossing multiple time zones quickly. When Dulles went to Egypt there was no such word as jet lag, and the word 'circadian' had yet to be invented.

Jet lag is the most obvious example of what can happen when the circadian system is disrupted and environmental and internal time are not properly aligned. Crossing more than three or four time zones in a jet aircraft uncouples these rhythms from the natural light/dark cycle and from each other. The various rhythms exhibit 'internal desynchrony' so metaphorically speaking the stomach ends up over Peking, the liver somewhere near Delhi, and the heart is still in San Francisco. It is even worse in space. Astronauts in orbit on the International Space Station may see sixteen dawns and dusks in 24 hours, so it is no wonder that sleeping tablets are the most frequently used medication in space.

Depending on how far you fly, the time of departure, and the direction in which you go, jet lag symptoms include: fatigue, insomnia, disorientation, headaches, and mood disturbances. Most people find it easier to cope with jet lag when flying from east to west, chasing the sun and experiencing a lengthening day. But overall the 'rule of thumb' is that it takes about a day for each time zone traversed for the circadian clock to adapt to the new local time, and practice does not help. Long-haul airline pilots who continually cross multiple times zones generally feel below par most of the time and this gets worse year after year. Pilot error, which constitutes a 'decision, action or inaction by a pilot of an aircraft determined to be a cause or contributing factor in an accident or incident', has often been attributed to jet lag. This seems to have been a contributing factor in the Air India plane that crash-landed in May 2010, killing 158 people. The official inquiry into the incident concluded that the captain was asleep for more than half of the three-hour flight and was 'disorientated' when he attempted to land the plane. Listening to the cockpit voice recorder, investigators heard 'heavy nasal snoring and breathing' from the pilot shortly before the crash.

Company executives who spend more time in the air than in the office never really adjust, even when they try to pretend that nothing has changed and resolutely refuse to reset their watches to local time. The number of time zones crossed increases the likelihood of incapacity and illness. American tennis players estimate they need to allow at least a week to get their body clocks back into gear when they fly from the USA to England for Wimbledon. A study from Israel even showed that the number of time zones crossed increased the risk of admission into a psychiatric hospital in tourists prone to mental illness.

The impact of jet lag has long been known by elite athletes, but the effects were first studied during the 1996 Olympic Games in Atlanta by Bjorn Lemmers. He measured the circadian rhythms of a group of German gymnasts before and during the games. These young men flew from Frankfurt to Atlanta, and they all suffered from jet lag with varying degrees of severity for up to one week after arrival. He monitored the athletes' blood pressure and measured body temperature, the saliva concentrations of cortisol and melatonin, as well as grip strength and overall jet lag symptoms. On the first day in Atlanta, the normal morning peak in cortisol occurred four hours earlier (closer to the time in Frankfurt). The normal rhythm in blood pressure, low at night versus high during the day, also took several days to adjust. Body temperature, which is tightly controlled by the circadian system and is normally highest at around 18.00–19.00 and is at its lowest around 04.00, was also about 4 hours phase advanced with local time, and training performances suffered. Eleven days after the time zone shift to Atlanta the rhythmic pattern in body temperature was still not fully adjusted in all athletes. In another study on the same German gymnasts, but this time flying eastwards to a competition in Osaka in Japan, jet lag symptoms lasted even longer. Again, physiology was abnormally phased, but in the opposite direction (delayed), compared to the local time zone.

These studies were on highly motivated, young men at their physical peak, yet even when superbly fit individuals fly across time zones there is a very prolonged disturbance of circadian-driven rhythmic physiology. Lemmers concluded that to be at the top of your game you need at least two weeks to recover from time zone transitions of 6 hours or more.

A group of sports fans at the Stanford University Sleep Disorders Clinic found that over a 25-year period, US West Coast American football teams beat East Coast teams 64 per cent of the time when the game was played at home. They concluded that the likely reason for this statistically significant result was circadian rhythm disruption. For the East Coast players travelling west, their internal time meant that games started at 21.00 and lasted until midnight; while for adapted West Coasters play began at 18.00 and was over by 21.00.

Horses also suffer from jet lag. Racehorses and show jumpers commonly compete around the world and are often flown across multiple time zones. Like humans, their performance suffers for a week or more after the flight, and trainers try to decrease the effects of jet lag by changing both feeding times and exercise schedules prior to the flight to correspond to those that will be used at the destination. The horses are also exposed to bright early morning light prior to an eastward journey as this helps to advance the body clock. Likewise, evening light is beneficial for horses travelling westward which will delay the clock (Chapter 4, Figure 15).

Even bees can get jet lag. In one experiment, French bees were loaded on to a jet and flown west to New York. When they landed they were still on Paris time and upon arrival emerged from their hives and went off searching for nectar when the flowers had yet to open.

Shift work

The misalignments that occur as a result of the occasional trans-meridian flight are transient. Shift working represents a chronic misalignment. Shift workers try to sleep during the day, but sleep is usually shorter and of poorer quality than when sleep occurs at night. They then work during the night at a time when the circadian system has prepared the body for sleep, and alertness and performance are low. In effect, they work when they are sleepy and sleep when they are not. Irrespective of how many years shift workers have been on a permanent night shift, nearly all (~97 per cent) night shift workers do not adjust to the nocturnal regime but are still synchronized to a diurnal rhythm because of light exposure.

Artificial light in the office or factory is dim compared to environmental light. Shortly after dawn natural light is some fifty times brighter than the 300 to 400 lux experienced in the workplace, and by noon natural light can be 250 times brighter. After leaving the night shift, an individual will usually experience bright natural light and the circadian system will always lock on to the brighter light signal as daytime and align internal biology to the diurnal state. In one study by Charles Czeisler and his team at Harvard University, night shift workers were exposed to 2,000 lux in the workplace and then completely shielded from any natural light during the day. Under these circumstances they became nocturnal. This, of course, is not a practical solution for most night shift workers.

SCRD, of the sort experienced by night shift workers, can lead to an increased risk of serious health conditions (Table 1). Short-term SCRD for a few days can have a big impact upon emotion and cognition, whereas longer term SCRD over years has been shown to increase the risk of cancer and cardiovascular disease.

Nurses are one of the best-studied groups of night shift workers. Years of shift work in these individuals has been associated with a broad range of health problems including type II diabetes, gastrointestinal disorders, and even breast and colorectal cancers. Cancer risk increases with the number of years of shift work, the frequency of rotating work schedules, and the number of hours per week working at night. The correlations are so strong that shift work is now officially classified as 'probably carcinogenic [Group 2A]' by the World Health Organization. Other studies of shift workers show increased heart and stroke problems, obesity, and depression. A study of over 3,000 people in southern France found that those who had worked some type of extended night shift work for ten or more years had much lower overall cognitive and memory scores than those who had never worked on the night shift.

SCRD also impairs glucose regulation and metabolism (Chapter 7). In 1999, Eve Van Cauter and her colleagues at the University of Chicago showed under laboratory conditions that sleep restriction in healthy young men led to signs of insulin resistance, which can ultimately lead to type II diabetes. Two gut hormones, leptin and ghrelin, seem to play a key role in this process. Leptin is produced by fat cells and is a signal of satiety; ghrelin is produced by the stomach and signals hunger, particularly for sugars. Together, these hormones regulate hunger and appetite. Van Cauter's team showed that restricting the sleep time of healthy young men in the lab for seven days caused their leptin levels to fall (~17 per cent) and their ghrelin levels to rise (~28 per cent), increasing their appetite, especially for fatty and sugary foods (increased by 35–40 per cent). Such a SCRD-induced distortion of appetite may explain why shift workers have a higher risk of weight gain, obesity, and type II diabetes. Significantly, night shift workers have elevated levels of the stress hormone cortisol, which has also been shown to suppress the action of insulin and raise blood glucose.

From these and many other observations it seems reasonable to conclude that both short-term and long-term night shift work carries varied and important health risks. But the studies undertaken to date are largely observational, and by their very nature cannot demonstrate a causal link between shift work and illness. Lifestyle factors associated with shift work may contribute to the increased health risk. For example, women who work nights have been found to have their children at a later age, eat less healthily, exercise less, drink more alcohol, and smoke more, all factors associated with weight gain, type II diabetes, and breast cancer. Understanding 'cause and effect' is critical, and the use of animal models to understand the mechanistic links between SCRD and poor health is very important. Recent studies on mice found that repeated changes in light schedules alone act as a causative factor for the development of cancer, weight gain, and other metabolic problems. While the studies were on rodents, the authors believe that their results provide compelling evidence that it is not lifestyle co-founding factors that are involved but circadian rhythm disruption itself. They argue strongly that individuals with hereditary (breast) cancer predispositions should not be exposed to any conditions that promote SCRD, such as shift work or frequent trans-meridian flight.

Although we have yet to establish mechanistically why shift work causes poor health, there is sufficient knowledge to try and mitigate some of the health issues associated with shift work. It seems both obvious and a moral duty that employers ensure that their shift workers have more frequent health screens; that they have access to an appropriate diet (low fats and sugars) in the workplace; and be supplied with electronic devices to warn individuals if they experience lapses in vigilance at work when using heavy equipment or on the drive home. In addition, the partners and families of night shift workers need to be aware that mood swings, loss of empathy, and irritability are common features of working at night.

Shift working and jet lag are excellent reminders that human existence is embedded within a dynamic 24-hour world. Until the

Table 1 Summary of the negative impact of sleep and circadian rhythm disruption (SCRD) associated with shift work on emotion, cognition, physiology, and health

Emotion	Cognition	Physiology and health
Increased:	*Impaired*:	*Increased risk of*:
Fluctuations in mood	Cognitive performance	Drowsiness
Depression and psychosis	Ability to multitask	Micro-sleeps
Irritability	Memory	Unintended sleep
Loss of empathy	Attention	Sensations of pain and cold
Frustration	Concentration	Cancer
Risk-taking and impulsivity	Communication	Metabolic abnormalities
Stimulant use (e.g. caffeine)	Decision making	Type II diabetes
Sedative use (e.g. alcohol)	Creativity	Cardiovascular disease
Illegal drug use	Productivity	Reduced immunity
Dissociated mental processing	Motor performance	Altered endocrine function

19th century, society was still largely agricultural and most people spent much of their time outdoors and lived their lives according to the natural day/night cycle. With increased industrialization, cheap electric light, and an increasingly 24/7 society, individuals have been detached from the solar cycle and the temporal order provided by the circadian system has been lost or diluted. The generations since the beginning of the 20th century have been living through a time of 'Great Circadian Disruption', and it is now becoming clear that this headlong rush into modernity, without a second thought to our biology, is having profound effects upon our health (see Table 1).

Circadian rhythm and sleep timing disorders

The physiology of the sleep/wake cycle involves the interaction between many brain regions and neurotransmitter systems (Chapter 6). In brief, wake/sleep can be considered as the interaction between a sleep 'reservoir' that is depleted during the day to produce a need for sleep often called 'sleep pressure', and a circadian drive that maintains wakefulness during the day and helps maintain sleep at night. Many of us choose, or are forced by economics, to work against our sleep/wake cycle and some of the consequences have been outlined in this chapter. However, for significant numbers of individuals SCRD arises from an endogenous breakdown of the complex physiological mechanisms driving and regulating sleep/wake. There are some seventy sleep disorders recognized by the medical community, of which four have been labelled as 'circadian rhythm sleep disorders', and in all of the conditions illustrated in Figure 7 there is evidence for an increased vulnerability to the health problems listed in Table 1.

The following circadian rhythm sleep disorders have been described to date:

(1) Advanced sleep phase disorder (ASPD), which is characterized by difficulty staying awake in the evening and difficulty staying asleep in the morning. Typically individuals go to bed and rise about three or more hours earlier than the societal norm. ASPD was first described in a large family from Utah. Some adult family members would go to bed early, around 19.00 to 20.00, and wake around 04.00. The family relationship demonstrated that there is a genetic component to the phenotype which has been pinpointed to a mutation substituting a serine with a glycine residue at a single amino

acid in one of the key genes of the molecular clockwork (Chapter 5).

(2) Delayed sleep phase disorder (DSPD) is a far more frequent condition and is characterized by a 3-hour delay or more in sleep onset and offset and is a sleep pattern often found in some adolescents and young adults. This often leads to greatly reduced sleep duration during the working week and extended sleep on free days. ASPD and DSPD can be considered as pathological extremes of morning or evening preferences (see Figure 6).

(3) Freerunning or non-24-hour sleep/wake rhythms occur in blind individuals who have either had their eyes completely removed or who have no neural connection from the retina to the brain. These people are not only visually blind but are also circadian blind. Because they have no means of detecting the synchronizing light signals they cannot reset their circadian rhythms, which freerun with a period of about 24 hours and 10 minutes. So, after six days, internal time is on average 1 hour behind environmental time.

(4) Irregular sleep timing has been observed in individuals who lack a circadian clock as a result of a tumour in their anterior hypothalamus (Chapter 4). Irregular sleep timing is commonly found in older people suffering from dementia. It is an extremely important condition because one of the major factors in caring for those with dementia is the exhaustion of the carers which is often a consequence of the poor sleep patterns of those for whom they are caring. Various protocols have been attempted in nursing homes using increased light in the day areas and darkness in the bedrooms to try and consolidate sleep. Such approaches have been very successful in some individuals, consolidating sleep and in parallel improving cognition.

Although insomnia is the commonly used term to describe sleep disruption, technically insomnia is not a 'circadian rhythm sleep

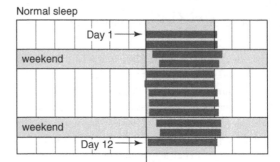

Normal sleep

Day 1 →

weekend

weekend

Day 12 →

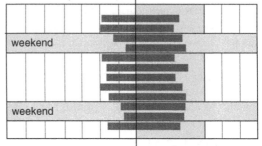

Advance sleep

weekend

weekend

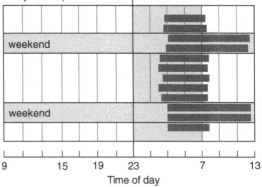

Delayed sleep

weekend

weekend

9   15   19   23   7   13

Time of day

7. Normal versus abnormal patterns in sleep timing. Black horizontal bars represent periods of sleep on consecutive work days and at the weekend. Weekends and night time are indicated.

Free-running sleep

Irregular sleep

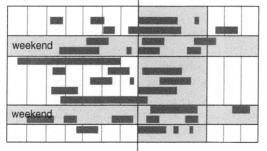

Insomnia

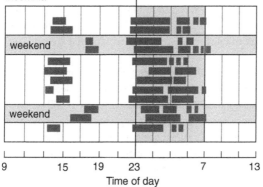

9   15   19   23   7   13

Time of day

disorder' but rather a general term used to describe irregular or disrupted sleep. This may manifest as reduced sleep (hyposomnia) or excessive sleep (hypersomnia). Insomnias probably arise from a complex interaction between the circadian timing system and other sleep drivers and modulators. Insomnia is described as a 'psychophysiological' condition, in which mental and behavioural factors play predisposing, precipitating, and perpetuating roles. The factors include anxiety about sleep, maladaptive sleep habits, and the possibility of an underlying vulnerability in the sleep-regulating mechanism. One-third of adults in Western countries experience difficulty with sleep initiation or maintenance at least once a week and suffer impaired daytime well-being and cognitive functioning.

Even normal 'healthy ageing' is associated with both circadian rhythm sleep disorders and insomnia. Both the generation and regulation of circadian rhythms have been shown to become less robust with age, with blunted amplitudes and abnormal phasing of key physiological processes such as core body temperature, metabolic processes, and hormone release. Part of the explanation may relate to a reduced light signal to the clock (Chapter 4). In the elderly, the photoreceptors of the eye are often exposed to less light because of the development of cataracts and other age-related eye disease. Both these factors have been correlated with increased SCRD. In addition, physical inactivity tends to limit natural light exposure, and the use of curtains and window shutters will decrease zeitgeber strength even further.

An age-related decline in clock function is correlated with a reduction in our cognitive abilities. Because sleep is important for memory consolidation (Chapter 6), SCRD will contribute to memory impairment. However, there are also likely to be additional age-associated neurodegenerative effects upon cognition. The mechanisms involved are again unclear, but the circadian system regulates processes that influence neurodegeneration, including oxidative stress responses, DNA damage repair mechanism, cellular cleaning process, and the removal of toxins. A loss of circadian control will increase rates of neurodegeneration. This will feed back and further disrupt circadian control, which will in turn promote age-related physiological decline. The result is a 'positive feedback loop' whereby circadian decline promotes physiological decline, which promotes neurological decline. In parallel, loss of circadian control will disrupt sleep. Efficient sleep promotes both memory consolidation and good mental health. If the relationships described here are correct, then preventing or mitigating SCRD during ageing will have a major impact across the health spectrum in the ageing population.

SCRD in mental illness

Mental illness is a vague term used to describe any behavioural pattern that causes either suffering or a reduced ability to function in ordinary life and encompasses conditions as diverse as anxiety, depression, mood disorders such as bipolar disorder, and psychotic conditions such as schizophrenia. SCRD is strongly associated with mental illness, and while most attention has been paid to mood disorders, especially seasonal affective disorder (SAD), SCRD is also prominent in the more severe, psychotic disorders such as schizophrenia.

SAD is a mood disorder in which individuals experience depressive symptoms in the winter but have normal mental health throughout the rest of the year. Symptoms include depression and lack of pleasure, difficulty waking, and a tendency to oversleep and overeat, especially carbohydrates, leading to weight gain. Other symptoms include a lack of energy, difficulty concentrating, and withdrawal from all social activities. SAD was first described in a 1984 paper by Norman Rosenthal and colleagues at the National Institute of Mental Health. Light therapy either before or during the depressive episode can either eliminate or reduce many of the symptoms of winter SAD. Light therapy using a light box exposes an individual to more than 2,000 lux, compared to normal domestic light levels of around 300 lux. Also exposure to sunlight by spending more time outside has shown to be effective. There are two favoured explanations for this action of light. The first suggests that supplementary light acts to entrain the circadian system in the winter months and so prevents internal desynchrony and the drift into SCRD; the second suggests that supplementary light will by some means increase serotonin levels throughout the brain, and elevated levels of serotonin are associated with feelings of well-being and happiness while low levels are linked to depression.

The relationship between schizophrenia and abnormal sleep was first described in the late 19th century by the German psychiatrist Emil Kraepelin. Today, SCRD is reported in more than 80 per cent of patients with schizophrenia, and is increasingly recognized as one of the most common features of the disorder. Sleep disturbances in schizophrenia are varied, with all of the abnormal patterns of sleep timing illustrated in Figure 7 observed in different individuals. The association between mental illness and SCRD was until recently considered to be driven by either social isolation, lack of employment, antipsychotic medication, or activation of the stress axis. Such a relationship between psychosis and SCRD is illustrated in Figure 8a. However, researchers at the Sleep and Circadian Neuroscience Institute (SCNi) at the University of Oxford have shown that SCRD in patients with schizophrenia persists independently of antipsychotic medication, social isolation, or lack of employment. These findings, along with the increased understanding of the neuroscience of sleep, suggest an alternative hypothesis; psychoses and SCRD share common and overlapping mechanistic pathways in the brain, as illustrated in Figure 8b.

Because the sleep/wake cycle involves an interaction between multiple brain regions and many neurotransmitter systems, abnormal neuronal circuitry that predisposes an individual to psychiatric illness is likely to have a parallel effect upon the

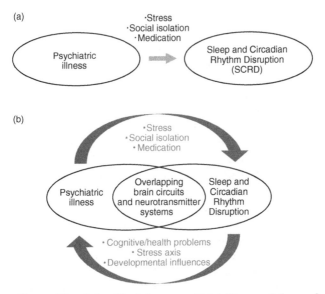

8. The possible relationships between psychiatric illness and sleep and circadian rhythm disruption (SCRD).

sleep/circadian system. Disruption of sleep will, likewise, impact upon many aspects of brain function, including activation of the stress axis, exacerbating or driving a range of health problems (Table 1), and in the young may have developmental consequences. Medication, substance abuse, social isolation, and/or activation of the stress axis associated with psychiatric illness will certainly impinge upon the sleep and circadian systems, but are shown in Figure 8b as contributors rather than drivers of SCRD in neuropsychiatric illness. Viewed in this context it is no surprise that SCRD is common across the mental illness spectrum, or that disruption of sleep/circadian biology might exacerbate a fragile mental health. Significantly, many of the health problems caused by SCRD (Table 1) are reported as co-morbid with neuropsychiatric illness, but in the clinic are rarely linked to the disruption of sleep.

The conceptual framework outlined in Figure 8b gives rise to four explicit predictions, that: (i) genes linked to mental illness will play a role in sleep and circadian rhythm generation and regulation; (ii) genes that generate and regulate sleep and circadian rhythms will play a role in mental health and illness. To date a surprisingly large number of genes have been identified that are associated with, and play a role in, *both* SCRD and mental illness; (iii) if the mental illness is not causative of SCRD, then SCRD may precede mental illness under some circumstances. Sleep abnormalities have indeed been identified in individuals prior to mental illness. SCRD precedes first episode or recurrent depression. Furthermore, individuals identified as 'at risk' of developing bipolar disorder and childhood-onset schizophrenia show SCRD prior to any clinical diagnosis of illness. Such findings raise the possibility that SCRD may be an important indicator (biomarker) in the early diagnosis of individuals with mental illness, and early diagnosis offers the possibility of early intervention; (iv) reduction of SCRD will have a positive impact upon mental health status. The Oxford team partially stabilized sleep in patients with schizophrenia who exhibited persistent persecutory delusions. Sleep stabilization was associated with a decrease in paranoid thinking along with a reduction in anxiety and depression. These data suggest that the stabilization of sleep timing can be an effective means to reduce the symptoms in a number of mental illnesses. Collectively the data emerging from Oxford and other institutes around the world support the general hypothesis that SCRD and mental illness are more accurately depicted and understood by the model depicted in Figure 8b rather than the original view shown in 8a.

SCRD in neurodegenerative disease

Neurodegenerative disease ultimately gives rise to abnormal/reduced neurotransmitter signalling that leads to defective brain function. So it is no surprise that SCRD is reported to be common in conditions such as Alzheimer's disease, and

Parkinson's disease. In Figure 8b 'Psychiatric illness' could be replaced with the term 'neurodegenerative disease' and the same relationships would apply. Neurodegenerative disease is usually progressive and irreversible, but sleep stabilization in these conditions is emerging as a possible approach to improve the overall health and quality of life (see Table 1) of patients, and in some cases even slow the progress of physical and mental decline.

The impact of fragmented night-time sleep in patients with Alzheimer's disease is very debilitating to both patients and caregivers and is a primary reason for patient institutionalization. The general neurodegenerative process in Alzheimer's disease almost certainly alters many aspects of sleep and circadian control. Brain nuclei in the anterior hypothalamus and the basal forebrain contain key regulatory circuits for sleep (Chapter 6), and degeneration of these regions in patients with Alzheimer's disease would seem likely to contribute to sleep problems. In addition, limited light exposure, which is often severe in institutionalized patients, will predispose individuals to freerunning sleep patterns (Figure 7). This will ultimately result in internal temporal desynchrony. Sleep stabilization in institutionalized patients with Alzheimer's disease improves night-time sleep along with higher daytime activity and less daytime sleep.

The earliest description of Parkinson's disease by James Parkinson (who was not only an apothecary and surgeon but also palaeontologist and political activist) included a reference to disturbed sleep. Today, it has been estimated that 80–90 per cent of patients with Parkinson's disease have disrupted sleep. Parkinson's disease is characterized by neuronal cell loss with Lewy body formation within the *substantia nigra* of the mid-brain leading to decreased brain dopamine. However, degeneration with Lewy body formation also occurs in the brainstem nuclei which are critical in sleep/wake regulation (Chapter 6). Non-Parkinson's diseases of the basal ganglia, such as Huntington's disease, also show the progressive development of SCRD along with atrophy in key structures involved in the regulation in sleep and circadian rhythms.

SCRD of the sort illustrated in Figure 7 has been reported in some individuals with multiple sclerosis (MS), suggesting that circadian and sleep structures suffer demyelination. Given the complex interactions of the multiple neuronal systems involved in sleep generation and sleep/wake control, demyelination in one or more of these neuronal systems could profoundly affect sleep and arousal in MS. In such patients sleep/circadian stabilization may improve quality of life, but no systematic studies have been undertaken to date.

If the relationships depicted in Figure 8b are broadly correct, then stabilization of the sleep/circadian system in mental illness and neurodegenerative disease would be expected to have a positive impact upon health. Evidence is emerging that such stabilization does indeed prove beneficial. For example bright-light phototherapy has been used as an entrainment signal for the circadian system and has been shown to alleviate some of the symptoms of several mood disorders including SAD, unipolar depression, and bipolar depression.

In addition to light, the pineal 'dark' hormone melatonin (Chapter 6) can also act partially to entrain the circadian system. In early studies, patients who had no eyes and were freerunning were given 3 mg of melatonin at bedtime and this moderately stabilized sleep/wake timing in some individuals. In January 2014, the US Food and Drug Administration approved the melatonin receptor agonist tasimelteon (trade name Hetlioz) as the first treatment for non-24-hour sleep/wake disorder following studies on 104 individuals with visual loss, but the effect was not dramatic. Treatment after several weeks increased night-time sleep and decreased daytime sleep duration in only 20 per cent of individuals compared with placebo controls. The extent to which

Table 2 Some approaches to achieve better sleep

Developing a good sleeping environment	Limiting stimulation before bed
Removing distractions from the sleeping space such as television and computers; make the sleeping space a place for sleep	Minimize light exposure at least 30 minutes before desired sleep time, and keep lights low in the sleeping space prior to sleep
Keep the sleeping space dark, not too warm, and quiet	Eat at the same time and finish the final meal of the day no later than three hours before desired bedtime
Invest in a comfortable bed, mattress, and pillows	Avoid caffeinated drinks in the afternoon and certainly before sleep
Try to maintain a consistent bedtime and wake-up time	Avoid the use of alcohol or other sedatives to induce sleep
If you wake, minimize light exposure, don't lie in bed frustrated but relax elsewhere and then return to bed	Engage in regular physical exercise, but not within 4 hours prior to going to sleep
Adopt a regular routine that you find particularly relaxing, such as reading or listening to music	Avoid stressful situations before sleep
If you cannot cope with your bed partner's snoring, find an alternative sleep space; this does not reflect upon your relationship	Avoid napping; if you nap then not for longer than 20 minutes and never 4 hours before sleep

melatonin or its agonists might be useful for sleep/wake stabilization in neurodegenerative disease and mental illness remains to be determined.

In combination with light and/or melatonin, social cues can also be useful in regulating the circadian/sleep system. Timed activities can influence daily patterns of light exposure and modify the timing of behaviour by associative learning and reinforcement. Meal timing, for example, can act as a strong stimulus for the synchronization of peripheral circadian rhythms in animals and humans, and could prove valuable when incorporated into cognitive behavioural therapy (CBT) paradigms. CBT aims to improve sleep habits and behaviours by identifying and changing the behaviours affecting the ability of an individual to sleep or sleep well. Such strategies are illustrated in Table 2. One key, but clinically unrecognized, element of CBT is to increase 'zeitgeber strength' by timing appropriate light exposure or meal times to the appropriate time of day (circadian phase).

Circadian rhythm research has mushroomed in the past twenty years, and has provided a much greater understanding of the impact of both imposed and illness-related SCRD. We now appreciate that our increasingly 24/7 society and social disregard for biological time is having a major impact upon our health. Understanding has also been gained about the relationship between SCRD and a spectrum of different illnesses. SCRD in illness is not simply the inconvenience of being unable to sleep at an appropriate time but is an agent that exacerbates or causes serious health problems. Developing better pharmacological, light, and CBT treatments for SCRD will have a global impact upon the economics of health care, and at an individual level improve the quality of life for countless patients and their caregivers.

Chapter 4
Shedding light on the clock

To be of any value a clock must be set to local time. If you were a resident of Greenwich in London during the 19th century you would have set your clock by a visual time signal from the observatory in the form of a 'Time Ball' on a tower that was dropped at precisely 13.00 every day. The first time ball was erected at Portsmouth, England, in 1829 by its inventor Robert Wauchope, a captain in the Royal Navy. However, when radio time signals were introduced in Britain from 1924, time balls became obsolete and many were demolished in the 1920s. Nowadays you would use the talking clock or the time on your smartphone, signals that originate from an atomic clock so accurate that if it had been started nearly 14 billion years ago at the Big Bang it would have only lost a second or two by now.

Most circadian clocks make use of a sun-based mechanism as the primary synchronizing (entraining) signal to lock the internal day to the astronomical day. For the better part of four billion years, dawn and dusk has been the main zeitgeber that allows entrainment. Circadian clocks are not exactly 24 hours. So to prevent daily patterns of activity and rest from drifting (freerunning) over time, light acts rather like the winder on a mechanical watch. If the clock is a few minutes fast or slow, turning the winder sets the clock back to the correct time.

Although light is the critical zeitgeber for much behaviour, and provides the overarching time signal for the circadian system of most organisms, it is important to stress that many, if not all cells within an organism possess the capacity to generate a circadian rhythm, and that these independent oscillators are regulated by a variety of different signals which, in turn, drive countless outputs (Chapter 7).

Colin Pittendrigh was one of the first to study entrainment, and what he found in *Drosophila* has been shown to be true across all organisms, including us. For example, if you keep *Drosophila*, or a mouse or bird, in constant darkness it will freerun. If you then expose the animal to a short pulse of light at different times the shifting (*phase shifting*) effects on the freerunning rhythm vary. Light pulses given when the clock 'thinks' it is daytime (subjective day) will have little effect on the clock. However, light falling during the first half of the subjective night causes the animal to delay the start of its activity the following day, while light exposure during the second half of the subjective night advances activity onset. Pittendrigh called this the 'phase response curve' or PRC, and Figure 9 illustrates how a PRC is deduced for a nocturnal animal such as a mouse.

In the upper part of Figure 9 (a–d) the light/dark cycle is shown and the dark line illustrates the duration of activity (also called 'alpha') on subsequent days. For the first days the animal is kept under a light/dark cycle of 12 hours of light and 12 hours of dark (LD 12:12). On day five, the lights were switched off and the animal was kept under constant darkness (DD), and it freeran with a period slightly shorter than 24 hours. To provide reference points under freerunning conditions, activity onset in a nocturnal animal is termed 'circadian time 12' or CT 12. The CT 0–12 is considered as the 'subjective day', and CT 12–24 is considered 'subjective night'. If the animal is exposed to a single 1-hour pulse of light during its subjective circadian day (□), as shown in

46

Figure 9 (a) there is usually no or little phase-shifting effect on the freerunning rhythm. This is called the 'dead zone'. At (b) the light pulse is given early in the subjective night; the effect is to start activity slightly later the next day (a delaying phase shift). In (c) the light exposure is later into the night and there is an increased delaying effect the following day. When light is given during the second half of the night (d), the effect is to advance the freerunning rhythm. If the phase shifts (a–d) are plotted against the circadian time the result produces a PRC.

Remarkably, the PRC of all organisms looks very similar, with light exposure around dusk and during the first half of the night causing a delay in activity the next day, while light during the second half of the night and around dawn generates an advance. The precise shape of the PRC varies between species. Some have large delays and small advances (typical of nocturnal species) while others have small delays and big advances (typical of diurnal species).

Light at dawn and dusk pushes and pulls the freerunning rhythm towards an exactly 24-hour cycle. In addition, the PRC also explains how activity is appropriately aligned to the expanding and contracting dawn/dusk signal across the seasons in non-equatorial zones. As illustrated in Figure 9, the size of the delaying phase shifts gets larger from subjective dusk into the night. So as night length gets shorter in the spring, delays will get bigger as more of the PRC is 'exposed' to light. This delaying effect is counterbalanced by larger advances as more of the PRC is exposed to light as dawn gets earlier. In nature, entrainment arises from the averaging of delays at dusk and advances around dawn. In some nocturnal animals, in northern latitudes exposure to the long days of spring and summer can greatly compress night-time activity, but at least this activity will occur mostly during the dark—the time of day which allows the animal the best chances of survival.

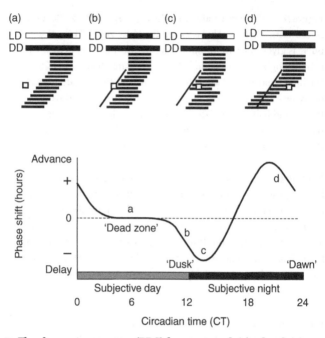

9. The phase response curve (PRC) for a nocturnal animal such as a mouse.

The easiest way to think about the delaying and advancing effects of light on the circadian system is to consider a nocturnal mouse out in the wild emerging from its burrow during early dusk. Assuming it does not get eaten, the mouse will get a pulse of light that will delay the clock, the activity pattern will start later the next day and the mouse will hopefully emerge after dusk and avoid the increased risk of predation. At the other end of the day, if the mouse has not retreated to its burrow around dawn then light will advance its clock and activity will occur earlier the next day, so it will have completed its foraging before dawn arrives. In this way the activity pattern of the mouse is constantly being pushed back and forth so that it self-corrects around dawn and dusk. The situation is the same for diurnal species except that activity patterns must be located during the day. Again, dusk light will delay and dawn light will advance the clock, centring activity to the day and not the night.

Light can act directly to modify behaviour. In nocturnal rodents such as mice, light encourages these animals to seek shelter, reduce activity, and even sleep, while in diurnal species light promotes alertness and vigilance. So circadian patterns of activity are not only entrained by dawn and dusk but also driven directly by light itself. This direct effect of light on activity has been called 'masking', and combines with the predictive action of the circadian system to restrict activity to that period of the light/dark cycle to which the organism has evolved and is optimally adapted.

The fruit fly, *Drosophila melanogaster*, along with rats and mice, has been a key model organism in circadian research. It is small, breeds quickly, has a well-characterized genome, is cheap to keep, and by using infrared light beams, which are interrupted by the flies' movements, measuring its circadian locomotion behaviour can be automated. *Drosophila* is also popular because although it is a fly, many of its genes are homologous to the vertebrates, including us. The adult *Drosophila* brain consists of approximately 100,000 neurons and around 150 in each brain hemisphere act as the 'master clock' of the fly. *Drosophila* possesses a variety of photoreceptors located in the eye, but also within the brain and in the clock cells themselves. This multiplicity of photoreceptors interact to communicate light information to the master clock.

Most invertebrates, like *Drosophila*, are translucent. As a consequence, all invertebrates appear to use photoreceptors within their brain to detect light directly for entrainment of their brain clocks, and although the eyes can contribute light information for entrainment they are not required. Further, many tissues of the invertebrates are directly light sensitive. Experiments have shown that isolated antennae, legs, or other organs from *Drosophila* maintained in culture solution will entrain circadian rhythms of gene expression in these organs to the light/dark cycle, and shift the rhythmic pattern of gene expression if the light/dark cycle is advanced or delayed.

Considerable amounts of light can also penetrate into the vertebrate brain. Pioneering studies from the 1900s to the 1970s by notable scientists including Karl von Frisch, Jacques Benoit, Eberhard Dodt, and Michael Menaker showed that birds, reptiles, amphibians, and fish (but not mammals) have 'extra-ocular' photoreceptors located within the pineal complex, hypothalamus, and other areas of the brain, and like the invertebrates, eye loss in many cases has little impact upon the ability of these animals to entrain. David Whitmore at University College London has shown that zebrafish possess photoreceptors in all of their tissues and use these photoreceptors to entrain local cellular clocks. The best candidate for this photosensitivity is a gene that encodes a light-sensitive molecule called 'teleost multiple tissue opsin', or TMT for short.

Mammals are strikingly different from all other vertebrates as they possess photoreceptor cells only within their eyes. Eye loss in all groups of mammals—the egg-laying platypus and echidna, the marsupials, and placental mammals—abolishes the capacity of these animals to entrain their circadian rhythms to the light/dark cycle. But astonishingly, the visual cells of the retina—the rods and cones—are not required for the detection of the dawn/dusk signal. There exists a third class of photoreceptor within the eye (see Figure 10).

Studies in the late 1990s by Russell Foster and his colleagues showed that mice lacking all their rod and cone photoreceptors could still regulate their circadian rhythms to light perfectly normally. But when the eyes were covered the ability to entrain was lost, so there had to be another photoreceptor within the eye. However, when this was originally proposed by Foster there was considerable hostility to the idea. Vision researchers argued that the eye had been the focus of intensive study for over 150 years,

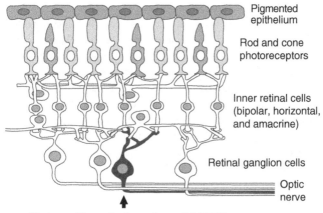

Pigmented epithelium

Rod and cone photoreceptors

Inner retinal cells (bipolar, horizontal, and amacrine)

Retinal ganglion cells

Optic nerve

Photosensitive retinal ganglion cell (pRGC)

10. Rods and cones convey visual information to the retinal ganglion cells via the second-order neurons of the inner retina—the bipolar and amacrine neurons. The optic nerve is formed from the axons of all the ganglion cells and this large nerve takes light information to the brain. A subset of photosensitive retinal ganglion cells (pRGC, shown as dark grey) also detect light directly.

how could an entire class of photoreceptor have been overlooked? Nevertheless, Foster's team persisted and work on the rodless/ coneless mouse, along with studies by David Berson in rats and Dennis Dacey and colleagues in the monkey, clearly demonstrated that the mammalian retina contains a small population of photosensitive retinal ganglion cells or pRGCs, which comprise approximately 1–2 per cent of all retinal ganglion cells (see Figure 10).

The pRGCs utilize the blue light-sensitive photopigment called 'melanopsin' or OPN4. The OPN4 gene was originally discovered by Ignacio Provencio, now based at the University of Virginia. He isolated this gene from the light-sensitive pigment cells or 'melanophores' located in the skin of the African clawed toad. Genetic ablation of the rods, cones, and melanopsin pRGCs in mice eliminates circadian responses to light, demonstrating that there are no additional photoreceptors either in the eye or body that regulate the circadian system. Although the rods and cones are not required for circadian entrainment, they are now known to contribute to the light responses of the melanopsin pRGCs under certain circumstances. Genetic silencing of OPN4 in the pRGCs does not block photoentrainment in mice. Mice can still entrain but with reduced sensitivities. Rods and cones send indirect projections via the inner retinal neurons (bipolar and amacrine cells) to the pRGCs (Figure 10), and it seems that in the absence of OPN4, the rods and cones can partially compensate for its loss. A complex pattern is emerging of how the different photoreceptor populations interact, and to complicate matters further, the eye itself has an independent circadian clock which changes the interaction strengths of the rods, cones, and pRGCs.

Just like mice, humans who have lost all of their rods and cones as a result of genetic disease show normal circadian entrainment and pRGC responses that peak in the blue part of the spectrum. This finding is having a major impact in the clinic. Ophthalmologists now appreciate that eye loss deprives us of both vision and a proper sense of time. Furthermore, genetic diseases that result in the loss of the rods and cones and cause visual blindness, often spare the pRGCs. Under these circumstances, individuals who have their eyes but are visually blind, yet possess functional pRGCs, need to be advised to seek out sufficient light to entrain their circadian system. The realization that the eye provides us with both our sense of space and our sense of time has redefined the diagnosis, treatment, and appreciation of human blindness.

The loss of extra-retinal photoreceptors in mammals is associated with their unique evolutionary history. The entire lineage of modern mammals arose from nocturnal ancestors. While the dinosaurs dominated the day, the mammals emerged at night. Because these ancient mammals emerged from their burrows at twilight, there would not have been sufficient light for long enough to be detected reliably by photoreceptors buried underneath layers of fur and a thick bony skull. As a result, mammals lost their extra-retinal photoreceptors and retained just retinal photoreceptors. From time to time, claims have been made that humans possess extra-retinal photoreceptors. However, such claims have never withstood rigorous scientific examination.

But where is 'the' circadian clock of mammals? Curt Richter at Johns Hopkins University in the 1950s and 1960s lesioned small parts of the rat brain in his hunt for the clock. He narrowed it down to somewhere deep in the brain, and almost certainly within the hypothalamus. Then in a series of experiments in the early 1970s, Robert Moore and Irving Zucker independently found what seemed to be 'the clock'. Knowing that circadian rhythms are entrained by the light/dark cycle they considered structures within the hypothalamus that received a direct retinal input. The suprachiasmatic nucleus, or SCN, is a pair of small nuclei located at the base of the hypothalamus that sit either side of the third ventricle just above the point where the optic nerves enter the brain—the optic chiasma. Large numbers of nerve fibres from the optic chiasma enter the SCN. Moore and Zucker lesioned the SCN and eliminated circadian rhythms (see Figure 11).

Moore and Zucker's work pinpointed the SCN as the likely neural locus of the light-entrainable circadian pacemaker in mammals (Figure 11), and a decade later this was confirmed by definitive experiments from Michael Menaker's laboratory undertaken at the University of Virginia. Small neural grafts from the SCN region of a mutant hamster with a short circadian period of 20 hours (*tau* mutant hamster) were transplanted into non-mutant hamsters whose own SCN had been lesioned and lacked 24-hour rhythms. This procedure not only restored circadian rhythms in wheel-running behaviour, but critically the restored rhythms always exhibited the period of the donor SCN regardless of the direction of the transplant; so a short 20-hour period SCN restored 20-hour rhythms when transplanted into a 24-hour

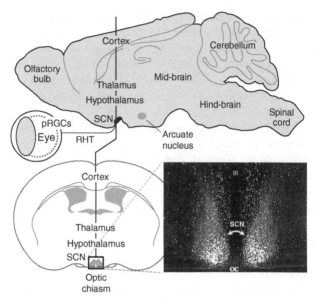

11. Diagram of the rat brain from the side showing the dedicated retino-hypothalamic tract (RHT) projection from the pRGCs of the eye to the suprachiasmatic nucleus (SCN) which contains the master circadian pacemaker of mammals. The frontal view of the brain shows the paired SCN located either side of the third ventricle (III) which sit on top of the optic chiasm (OC). The enlarged picture of the SCN shows individual SCN neurons (white dots) and projections from these 'clock cells' to the thalamus.

hamster, and a 24-hour SCN transplanted into a 20-hour animal restored 24-hour rhythms. These experiments established the SCN as the 'master circadian pacemaker' of mammals. This was a transformative discovery, and the study of circadian rhythms in mammals shifted from phenomenology to neuroscience.

The SCN itself is made up of autonomous, single-cell circadian oscillators sufficiently stable to generate circadian rhythms of neuronal firing for at least six weeks in isolation (*in vitro*)—each neuron is a proverbial clock in a dish. This was first shown in a critical experiment in the mid-1990s by Steven Reppert and colleagues. They dispersed the SCN of neonatal rats into individual cells which were fixed on to a grid of microelectrodes. They then recorded spontaneous action potentials (spikes) from individual neurons for days and weeks. Individual neurons displayed prominent circadian rhythms in firing rate, but the phases of these individual rhythms were all different, showing that SCN neurons act as individual clocks and that the basic oscillation lay within individual cells and was not the emergent property of a network of cells.

There are around 20,000 or so neurons in the mouse SCN, but they are not identical. Some receive light information from the pRGCs and pass this information on to other SCN neurons, while others project to the thalamus and other regions of the brain, and collectively these neurons secrete more than one hundred different neurotransmitters, neuropeptides, cytokines, and growth factors. The SCN itself is composed of several regions or clusters of neurons, which have different jobs. Furthermore, there is considerable variability in the oscillations of the individual cells, ranging from 21.25 to 26.25 hours. Although the individual cells in the SCN have their own clockwork mechanisms with varying periods, the cell autonomous oscillations in neural activity are synchronized at the system level within the SCN, providing a coherent near 24-hour signal to the rest of the mammal.

We are still unclear about the role of the specific cell types in the SCN but one group of neurons designated as 'Nms cells' are of particular importance. They produce a neuropeptide known as neuromedin S (NMS) and are found mainly in the central region of the SCN. In a series of sophisticated experiments using transgenic mice, Joe Takahashi's group at the University of Texas showed that lengthening the intracellular circadian period of Nms neurons lengthens the animal's behavioural circadian period, whereas abolishing the circadian rhythmicity of Nms neurons or blocking synaptic transmission from Nms neurons leads to the loss of coherent circadian rhythms across the SCN. In short, the Nms neurons control SCN network synchrony through intercellular synaptic transmission. Such a mechanism generates a precise, sustained, and robust circadian rhythm in the SCN, and while other cells in a variety of tissues and organs can generate circadian rhythms they lack this sustainability and precision. Indeed, the SCN rhythm is so robust that when brain slices that include the SCN are maintained in culture, circadian rhythms can persist for well over a year.

SCN neurons exhibit a circadian rhythm of spontaneous action potentials (SAPs), with higher frequency during the daytime than the night which in turn drives many rhythmic changes by alternating stimulatory and inhibitory inputs to the appropriate target neurons in the brain and neuroendocrine systems. The daytime frequency of SAPs is higher during the day in both diurnal and nocturnal mammals, which means that the 'nocturnal versus diurnal switch' that drives the behavioural differences between night active and day active animals must be located downstream from the SCN. The SCN projects directly to thirty-five brain regions, mostly located in the hypothalamus, and particularly those regions of the hypothalamus that regulate hormone release. Indeed, many pituitary hormones, such as cortisol, are under tight circadian control. Furthermore, the SCN regulates the activity of the autonomic nervous system, which in turn places multiple aspects of physiology, including the sensitivity of target tissues to hormonal signals, under circadian control.

In addition to these direct neuronal connections, the SCN communicates to the rest of the body using diffusible chemical signals. In the 1990s, Rae Silver and her colleagues at Columbia University transplanted the SCN contained within tiny semi-permeable capsules into SCN-lesioned animals. The capsule prevented neural connections being re-established but allowed chemical communication from the transplanted SCN to diffuse out. Even without a neural connection, some circadian rhythms were restored. Vasopressin (VP) has been considered a strong candidate for this chemical communication as the SCN has a large population of VP-containing neurons and there is a marked day/night rhythm of the peptide in the cerebrospinal fluid of a variety of mammals. Furthermore, robust circadian rhythms of VP secretion occur when SCN brain slices are maintained in culture. However, a clear link between rhythms in VP release from the SCN and rhythmic physiology and behaviour has not yet been established.

The SCN is the master clock in mammals but it is not the only clock. There are liver clocks, muscle clocks, pancreas clocks, adipose tissue clocks, and clocks of some sort in every organ and tissue examined to date. While lesioning of the SCN disrupts global behavioural rhythms such as locomotor activity, the disruption of clock function within just the liver or lung leads to circadian disorder that is confined to the targeted organ. In tissue culture, liver, heart, lung, skeletal muscle, and other organ tissues such as mammary glands express circadian rhythms, but these rhythms dampen and disappear after only a few cycles. This occurs because some individual clock cells lose rhythmicity, but more commonly because the individual cellular clocks become uncoupled from each other. The cells continue to tick, but all at different phases so that an overall 24-hour rhythm within the tissue or organ is lost.

The discovery that virtually all cells of the body have clocks was one of the big surprises in circadian rhythms research. Ueli Schibler led the team at the University of Geneva that discovered that mouse fibroblast cells (important cells in connective tissues) that had been cultured for a long time could still be 'shocked' into displaying circadian rhythms by introducing a single dose of blood serum to the culture medium. This in some way synchronized the individual circadian rhythms of the fibroblasts to the same circadian phase, and so the entire population of fibroblasts showed a synchronous rhythm that could then be detected.

Schibler's discovery led to the appreciation that the SCN, entrained by pRGCs, acts as a pacemaker to coordinate, but not drive, the circadian activity of billions of individual peripheral circadian oscillators throughout the tissues and organs of the body. The signalling pathways used by the SCN to phase-entrain peripheral clocks are still uncertain, but we know that the SCN does not send out trillions of separate signals around the body that target specific cellular clocks. Rather there seems to be a limited number of neuronal and humoral signals which entrain peripheral clocks that in turn time their local physiology and gene expression. The SCN also receives feedback signals from the periphery that allow the whole body to function in synchrony with the varying demands of the 24-hour light/dark cycle. In addition, the circadian rhythms of many organs and tissues are themselves interconnected, providing further coherence to the circadian network that regulates physiology and behaviour. However, there are occasions when the SCN and the peripheral clocks may uncouple from each other.

The behaviours of rats are strongly regulated by their circadian rhythms, not least feeding rhythms. When given unlimited food they eat most of their food during the night when they are active. However, Michael Menaker's laboratory showed that when food availability is restricted to only a few hours during the light phase each day, when they would normally be asleep, they quickly alter their circadian rhythms of physiology and behaviour. Locomotor activity, body temperature, corticosterone secretion, and metabolic systems realign to the timed food availability. However, the clockwork in the SCN remains aligned normally to the light/dark cycle. Under these circumstances, the clocks within the liver, gut, and other organs detach themselves from the SCN and are shifted to the availability of the food.

Rats can predict this regular arrival of food and show an increase in activity shortly before the time when the food is expected. This is called food-anticipatory activity (FAA). Because the rats showed anticipation, this provided overwhelming support to the conclusion that FAA must be generated by a self-sustained timing mechanism. Furthermore, and surprisingly at the time, this anticipatory activity was shown to occur in SCN-lesioned animals maintained on a 24-hour feeding schedule, demonstrating the existence of an SCN-independent food-entrainable oscillator (FEO).

Many organs seem to possess an FEO. A key mechanism whereby food acts as a zeitgeber on these FEOs seems to involve the release of hormones including leptin and ghrelin. When the stomach is empty ghrelin is secreted, but when the stomach is stretched secretion stops. Ghrelin acts on hypothalamic brain cells to increase hunger, increasing gastric acid secretion and gastrointestinal motility to prepare the body for food intake. Humans given injections of ghrelin feel voraciously hungry and eat more. Leptin is made by fat (adipose) cells and opposes the action of ghrelin by inhibiting hunger. Both ghrelin and leptin are regulated by food availability and act on receptors in the arcuate nucleus of the hypothalamus (Figure 11) to regulate appetite to achieve energy homeostasis, but multiple cell types outside the brain have leptin and ghrelin receptors. As a result these hormones could effectively provide global signals to peripheral clocks around the body signalling the absence or availability of food (Chapter 7).

From an adaptive point of view, FEOs allow an organism to uncouple from the light-entrained SCN and take advantage of periodic food availability at a time normally associated with sleep. The FEO can activate arousal, appetite, digestive secretions, and metabolism just before receiving food, allowing the animal to take full advantage of a predictable food source that may occur at the 'wrong' time of day. In the natural world an animal will have to balance the drive from a FEO that promotes opportunistic behaviours and the exploitation of a novel food resource, with the increased exposure and a greater chance of predation.

In addition to light and food, there are yet more zeitgebers such as physical exercise and temperature. In rodents, scheduled physical activity, such as wheel-running during the day, will phase shift and entrain circadian rhythms of locomotor behaviour, while in humans several studies have reported the phase-shifting effects of physical exercise. A single bout of physical exercise at night was demonstrated to phase-delay the circadian rhythms, while scheduled physical exercise during the waking period can help achieve stable entrainment. Periodic temperature cycles with periods around 24 hours, even with a temperature range of 1–2°C, can entrain all ectothermic (poikilothermic) organisms (insects, fish, reptiles etc.), but have very little or no impact upon homeotherms such as birds and mammals. Although isolated and maintained in culture, mammalian cells can be entrained to a temperature cycle. Temperature changes probably act directly on the clocks of ectotherms and mammalian cells in culture by altering membrane properties, ion homeostasis, calcium influx, and other signal cascades, and in this regard resemble some of the signalling pathways activated by light (Chapter 5).

The circadian system in mammals exhibits a remarkable adaptive plasticity, enabling animals to concentrate foraging efforts at times of day when food is most likely to be found, and to adjust physiological and metabolic rhythms to optimize the absorption, utilization, and storage of ingested nutrients. The SCN is the light-entrained oscillator (LEO) regulating the sleep/wake cycle and allied behaviours while the FEOs in the peripheral cells are entrained to feeding patterns. When the LEO is weakened (e.g. by a weak or non-24-hour light signal), then the action of the FEOs is enhanced. Conversely, disruption of FEOs (e.g. unpredictable or constant food availability) enhances SCN drivers. After the LEO and FEO are forced out of synchrony, the SCN influence is overridden to allow the animal to avoid starvation. Surprisingly, and despite considerable effort, the location of the FEO has not been definitively established.

Clearly the early mammalian model of a hierarchical circadian system with the SCN alone dictating circadian rhythmicity was far too simplistic. The circadian timing system is turning out to be a complex and highly sophisticated method of handling an organism's temporal alignment with both its external and internal environment. In parallel with this new understanding of how the circadian system is organized, there has been equally remarkable progress in the understanding of how a single cell can generate an endogenous 24-hour circadian oscillation that can be regulated by external zeitgebers.

Chapter 5
The tick-tock of
the molecular clock

In August 1765, a group of six experts gathered at the house in London of John Harrison to examine a clock. They took the back off and looked at the mechanism. A week later they certified that they understood fully its inner workings. H5, as the clock was called, won the Longitude Prize for Harrison and revolutionized navigation at sea. Harrison's clock translated the kinetic energy used to wind it into stored potential energy in the spring that was then released in a precisely regulated manner via an escapement mechanism to produce a rhythm. A gear train converted this beat into hands showing minutes and hours. It is a work of precision mechanical engineering, but the principle of establishing and maintaining a persistent and robust oscillation, however achieved, is at the heart of every clock, whether it is a fantastically accurate atomic clock, H5, or the rhythmic circadian clocks found across organisms as diverse as mammals, birds, reptiles, fish, insects, plants, fungi, algae, bacteria, and even the archaeon Halobacterium.

Until 1953 it was not remotely possible to examine the workings of the circadian clock by metaphorically taking the back off it. Both the conceptual framework and the technology were absent. The leap forward came after Crick and Watson's publication on the structure of DNA. Their paper and the burgeoning techniques of molecular biology have had a huge impact across all biology, and not least on our understanding of circadian rhythms.

Seymour Benzer was one of the pioneers. Benzer started as a physicist, but turned to biology and made his mark by laboriously fine-mapping the genes in a bacteriophage (a virus that infects and replicates within a bacterium). After ten years of sequencing virus genes, his interests switched to studying the genetic basis of behaviour. He was struck that his two daughters had such different personalities and in his own words 'I got interested in this general problem of personality and behaviour—how much is genetics and how much is environment? And how do you study such a problem?'

Based at the California Institute of Technology, Benzer chose to study *Drosophila melanogaster*. This tiny fruit fly had been instrumental in the development of genetics through much of the 20th century and it had been well established that different physical forms (phenotypes) of *Drosophila* with descriptive names like 'curly wings' were the result of changes in single genes. Benzer reasoned that specific behaviours might also be influenced by the action of a single gene. It seems incredible now, but fifty or so years ago biologists were split between those who agreed with Benzer's basic point that it would be possible to show that single genes influence behaviour and those who said he was absurd, because behaviour was far too complex to be defined by the influence of a single gene.

Ron Konopka, a PhD student in the group, was fascinated by the emerging field of circadian rhythms. In 1970, he and Benzer decided to try and find mutant flies with an altered circadian property. Konopka used both eclosion (emergence from the pupal case) and locomotor rhythms as his chosen circadian measures. Adult *Drosophila* eclose from their pupal cases during the early morning. Pupae kept on a regular schedule of 12 hours of light followed by 12 hours of darkness, and then subjected to constant darkness, will emerge at the time of day when they expect dawn (subjective dawn). In the wild this is the time when humidity is highest, which allows the flies to pump out their wings and to harden their cuticle before it gets too dry in the afternoon. This explains the name *Drosophila*, derived from the Greek for 'dew-lover'. This eclosion timing is effectively 'gated' so if the adult does not emerge at dawn on the first day when it is ready to go, then the fly will wait another 24 hours before it emerges. In parallel Konopka and Benzer studied the individual locomotor activity of flies in a glass tube, monitoring their movement by the interruption of an infrared light beam in darkness.

Konopka treated male flies with the mutagen ethyl methanesulfonate to induce point mutations in their DNA, and mated them with non-mutagenized females, and examined the eclosion timing and locomotor rhythms of the offspring. After 200 tries he found three types of mutant flies in which the circadian period had been profoundly altered. The three mutants he described were: a mutant that emerged/eclosed late and had a long freerunning period of about 28 hours; a mutant that emerged/eclosed early and had a short freerunning period of about 19 hours; and a mutant with no obvious rhythm at all (arrhythmic). All three mutant strains were mapped to the same locus on the X-chromosome. He called this mutated gene *period* (*per*) and the mutant alleles *per*L (for long period), *per*S (for short), and *per*0 (for the arrhythmic variant). Not only had Konopka demonstrated that at least one clock gene must exist but he had also identified by mutation the first behavioural gene in any organism!

After completing his PhD, Konopka undertook a postdoc with Colin Pittendrigh, who had done so much to provide a formal understanding of circadian rhythms, and learned the details of circadian biology before returning as an assistant professor to Caltech. Despite this pivotal work, sadly it did not turn out well for Konopka. During the 1970s he published a few papers, but these were not sufficient to give him tenure; eventually he left California and moved to Clarkson College in upstate New York, but again was denied tenure because of a lack of productivity and left academia in the late 1980s. He returned to California, lived in a trailer, had a difficult personal life, and worked as a computer consultant. He died in 2015, but is remembered fondly and respected greatly by the circadian community. The molecular study of chronobiology began with Konopka's PhD work with Benzer. They had taken the back off the circadian clock to examine its workings and this was the first step along the long road to deducing how a biological clock is built—asking what are the molecular analogues of gears and cogs, how do they mesh, what regulates them, and how do they establish a molecular cycle with circadian characteristics?

Circadian clocks in animals and plants arise from multiple and interconnected transcription–translation feedback loops (TTFLs) that ultimately ensure the proper oscillation of maybe thousands of genes in a tissue-specific manner. Which prompts the question, what is a TTFL?

Genetic information is held within DNA. This information has to be converted or transcribed into an mRNA copy before it can be used as a template for the synthesis of a corresponding protein. The sequence of bases in the mRNA strand mirrors the sequence of bases on the DNA strand. The strand of mRNA then leaves the nucleus in eukaryotes (organisms whose DNA is in the form of chromosomes contained within a distinct nucleus) and enters the cytoplasm of the cell where the mRNA interacts with ribosomes which are the 'protein assembly factories'. The information is 'translated' when transfer RNAs line up the amino acids, codon by codon, along the mRNA strand to generate a protein.

The sequence of DNA that provides the information to make a protein can be thought of as having two parts, a coding region and a promoter or regulatory region. The coding region contains the sequence of bases (codons) that defines the amino acid sequence of the proteins, whereas the promoter region(s) consists of a sequence of bases to which 'transcriptional regulators' bind. Many transcriptional regulators have a DNA-binding bHLH (basic helix-loop-helix) motif that acts to switch on and off the coding region of a gene and initiate transcription. Multiple transcription factors may be involved in driving the transcription of a single gene. Significantly, in some cases the protein of a gene made in the cytoplasm may then enter the nucleus and inhibit the transcriptional regulators driving the expression of its own gene—a classic negative feedback loop. When the protein is degraded by cellular processes, transcription can occur once again.

Three individuals have propelled our understanding of the *Drosophila* molecular clock forwards: Jeffrey Hall, who was another postdoctoral student of Benzer's who then moved to Brandeis University; Michael Rosbash, already at Brandeis; and Michael Young at Rockefeller University. They worked independently, and sometimes together, and succeeded in cloning the *Drosophila per* gene by the mid-1980s. The next significant step came from Hall and Rosbash who showed that PER, the protein transcribed from the *per* gene (the convention is that genes are written in *italic* and the protein they express in capitals), cycled in a small number of 'lateral neurons' within the fly brain. Under constant conditions there was a high-amplitude 24-hour *per* gene mRNA cycle in the heads of flies which peaked early in the subjective night and several hours before PER protein reached its peak.

In the early 1990s, Hall and Rosbash's laboratories showed that over several hours PER accumulates in the cytoplasm, and as the concentration of PER increased it was transported into the nucleus, peaking in the middle of the night, approximately 6–8 hours after the peak of *per* mRNA. PER protein would then override the positive (+) drive of transcriptional regulators and inhibit (−) its own gene. This reduced transcription, and ultimately translation, of new PER protein. Existing PER would then be degraded in both the nucleus and cytoplasm. With little PER protein, the inhibitory action of PER on the transcriptional drive of the *per* gene ceases and fresh *per* mRNA is transcribed.

This negative feedback loop would repeat indefinitely, taking approximately 24 hours for each cycle (Figure 12).

The concept was prescient but there were problems in the details. The PER molecule lacked the bHLH sequence motif that would enable it to bind directly to the DNA promoter region and hence function as an inhibitory transcriptional regulator—inhibiting its own transcription. The model was stuck until Young discovered another clock gene, the *timeless* gene (*tim*) in *Drosophila* in 1994. *Tim* mRNA was shown to cycle at about the same time as *per*, and critically TIM and PER protein were shown to bind together to form a complex. But as neither PER nor TIM nor the PER-TIM dimer bind to DNA, how could they regulate their own genes? The answer to this puzzle and how to close the feedback loop came in 1998.

Using mutagenesis, Hall and Rosbash identified two mutants which were arrhythmic, one called *Clockjrk* and the other, *cycle0*. The corresponding proteins, CLOCK and CYCLE, encoded bHLH regions that could bind *per* and *tim* DNA. In addition they both had a domain called PAS which allowed them to bind together as the CLOCK-CYCLE dimer. PER also had a PAS domain, and so now, one could see how the PER-TIM dimer might regulate the CLOCK-CYCLE dimer through the mutually shared PAS domains. Thus CLOCK-CYCLE are the transcription factors that activate the *per* and *tim* genes, and PER and TIM are the negative autoregulators which interact with CLOCK-CYCLE via the PAS region. The door was open to develop the current description of the *Drosophila* molecular clock which is shown in detail in Figure 13. At first sight the diagram is daunting, and if you are not interested in the details skip Figure 13 and the accompanying description, but it is worth making the effort as this is one of the great achievements in understanding how the interaction of genes and their protein products can ultimately give rise to physiology and behaviour.

The *Drosophila* clock mechanism consists of three TTFLs (Figure 13). The PER-TIM loop is the central or primary TTFL

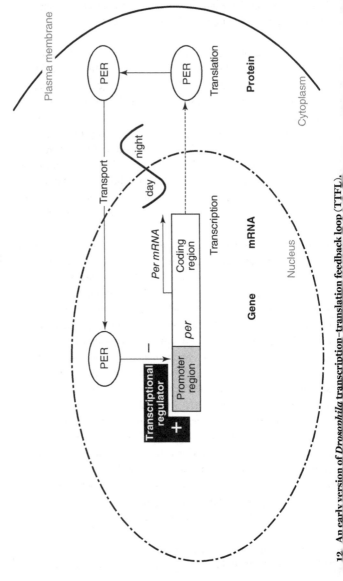

12. An early version of *Drosophila* transcription–translation feedback loop (TTFL).

and the secondary and tertiary loops help to stabilize the primary TTFL. The core feedback loop of the circadian TTFL is initiated when a transcription factor complex composed of CLOCK (CLK) and its bHLH partner CYCLE (CYC) bind to E-box (Enhancer-box) sequences in the *per* and *tim* promoters and activate transcription.

PER and TIM proteins start to accumulate around dusk, with TIM levels rising more slowly, and both peak late at night. PER monomers are phosphorylated by the kinase DBT (doubletime) leading to PER degradation, so PER fails to accumulate early at night. However, as TIM levels rise, TIM binds to the PER-DBT complex, effectively protecting PER from further significant phosphorylation and degradation.

Within the complex of TIM-PER-DBT, TIM and PER are phosphorylated by other kinases (SHAGGY/SGG and CASEIN KINASE 2/CK2). This phosphorylation allows the TIM-PER-DBT complex to move into the nucleus late at night. Once in the nucleus this complex binds to CLK-CYC and inhibits their transcriptional drive. The TIM-PER-DBT complex is then degraded by two mechanisms: (i) PER is released from the complex to be further phosphorylated in the nucleus by DBT which leads to the sluggish break-up of the TIM-PER-DBT complex. This slowly releases CLK-CYC from inhibition during the first half of the day and this permits *per/tim* transcription to resume; (ii) light at dawn triggers the degradation of TIM. While TIM itself is not responsive to light, the protein product of the *cryptochrome* (*cry*) gene, CRY, is activated by blue light which causes it to change shape and bind to TIM. This CRY-TIM binding can occur in either the cytoplasm or the nucleus (depending on when a light stimulus is applied), and TIM is then phosphorylated by yet another kinase and targeted for degradation by the F-box protein jet lag (JET).

Other photoreceptor molecules and mechanisms also trigger TIM degradation but their actions on TIM are poorly understood.

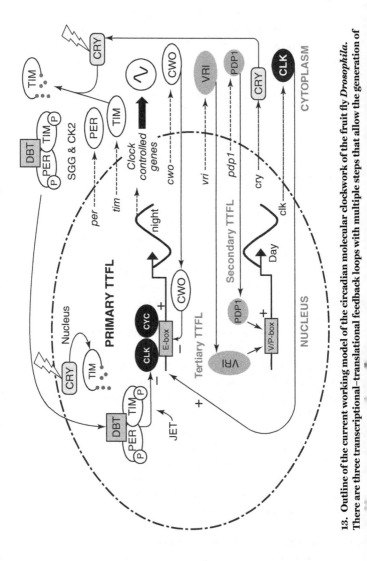

13. Outline of the current working model of the circadian molecular clockwork of the fruit fly *Drosophila*. There are three transcriptional–translational feedback loops with multiple steps that allow the generation of

So when morning light arrives TIM, within the TIM-PER-DBT complex, is degraded and as a result PER then becomes 'de-protected' and is rapidly phosphorylated by DBT in the nucleus. PER is then degraded and then the whole TIM-PER-DBT complex collapses allowing CLK-CYC mediated transcription to occur once again.

The consequences of CLK-CYC binding on to the E-box promoters of three additional genes: *vrille (vri)*; *par domain protein 1 (pdp1)*; and *clockwork orange (cwo)*, results in the formation of two more TTFLs within the molecular clockwork. The secondary TTFL involves the action of VRI and PDP1 on CLK transcription, which is effectively driven down overnight. VRI accumulates rapidly and binds VRI/PDP1-boxes (V/P-box) in the *Clk* promoter and represses *Clk* transcription, while another transcription factor PDP1 is involved in driving *Clk* transcription. In this way VRI and PDP1 act antagonistically to time CLK production so that it rises towards the early hours of the morning and peaks during the first half of the day. The *cry* gene is also under clock control such that *cry* mRNA cycles and peaks at dawn. However, the CRY protein can only accumulate at night (because light degrades it), so it peaks at the end of the night and is thus ready to promote the degradation of TIM by light at dawn.

The third (tertiary) TTFL is formed by the clockwork orange protein (CWO). In this feedback loop CLK-CYC activate *cwo* transcription. CWO then feeds back to inhibit CLK-CYC transcription and CWO expression. In this manner CWO reinforces the action of TIM-PER-DBT inhibition of CLK-CYC transcription.

Finally, the presence of an E-box in the promoter of very many genes means that they can also be under circadian control. These clock-controlled genes (*ccgs*) are not involved in the core clock mechanism but control overt rhythmicity directly or indirectly through multiple downstream mechanisms.

This complex arrangement is needed because from the moment a gene is switched on, transcription and translation usually takes two hours at most. As a result, substantial delays must be imposed at different stages to produce a near 24-hour oscillation. As shown in Figure 13, the generation of a robust circadian rhythm that can be entrained by the environment is achieved via multiple elements, including the rate of transcription, translation, protein complex assembly, phosphorylation, other post-translation modification events, movement into the nucleus, transcriptional inhibition, and protein degradation.

Another key issue is how can this molecular mechanism achieve temperature compensation? One theory is that temperature compensation is an emergent property of a molecular network. It is envisaged that the reactions of the constituent parts of the molecular clock are temperature-sensitive in the usual way. However, the quickening of some reactions such as transcription and translation will shorten the period of the clock, while the speeding up of other reactions such as protein degradation in the cytoplasm will lengthen the period of the clock. The lengthening of some processes is cancelled out by a shortening of other processes, with an overall effect close to zero.

Light resetting in the fly ultimately gives rise to TIM degradation (Figure 13). But resetting circadian behaviour via light also requires neural interactions. *Drosophila* is crepuscular, meaning that it is at its most active at dawn and dusk. The 150 neurons of the lateral neurons that express clock genes are arranged in seven distinct clusters. Michael Rosbash, in a series of elegant experiments, concluded that the ventral lateral neurons (LNvs) generated the morning peak in locomotor activity (the morning or M cells) whereas the dorsal lateral neurons (LNds), probably with some other dorsal clock neurons (the DNs) controlled the evening peak in locomotor behaviour. But the E and M cells talk to each other. Through genetic manipulation, M cells can be made to run with a faster circadian cycle, and this alters the timing of the E (dusk) component of behaviour, showing clearly that signals from M cells to E cells help achieve entrainment. Later experiments then showed that in winter, with long nights, the M cells dominate the circadian network, whereas in the summer the E cells perform this function.

Remarkably similar mechanisms have been found in mammals. Although the molecular players may differ from *Drosophila* and mice, and indeed even between different insects, the underlying principles apply across the spectrum of animal life. Joe Takahashi's research group, now based at the University of Texas, has been one of the key pioneers in dissecting the mammalian molecular clockwork and he, like Konopka with *Drosophila*, started by exposing mice to mutagens. Using this approach, Takahashi and colleagues isolated mice with an abnormal circadian period and discovered the mammalian *clock* gene (*Circadian Locomotor Output Cycles Kaput*) in 1994 and cloned the gene in 1997. Homologs of the mammalian *clock* gene were then subsequently isolated in *Drosophila*. In addition, Takahashi cloned the mutant *tau* gene identified in 1988 by Michael Menaker and Martin Ralph in hamsters at the University of Virginia. This gene turned out to be *Casein Kinase 1 epsilon* (*CK1e*) and it is closely related to the *doubletime* (*dbt*) gene/protein of *Drosophila*. Both DBT (*Drosophila*) and CK1 (mouse) interact with and regulate PER levels. These key insights led to the development of our current understanding of the molecular clockwork of the mouse, which is illustrated in detail in Figure 14.

The dynamic force of the mammalian molecular clockwork is the transcriptional drive provided by two proteins named CLOCK (CLK), which heterodimerizes with 'Brain muscle arnt-like 1' (BMAL1; Figure 14). Both are orthologous to *Drosophila* CLK and CYC respectively (Figure 13).

The *Bmal1* gene transcription produces a rhythmically produced BMAL protein that heterodimerizes with a constitutively

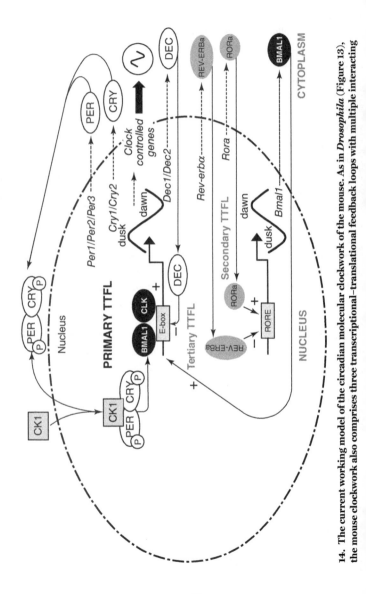

14. The current working model of the circadian molecular clockwork of the mouse. As in *Drosophila* (Figure 13), the mouse clockwork also comprises three transcriptional–translational feedback loops with multiple interacting

expressed CLK. The CLK-BMAL1 complex binds to E-box promoters driving rhythmic transcription of the *per 1–3* and two cryptochrome genes (*Cry1, Cry2*). Unlike *Drosophila*, the mammalian CRY proteins are not light sensitive, as was originally assumed, and do not act in the entrainment pathway. Instead, the CRYs functionally replace *Drosophila* TIM to associate with the PERs (PER1, PER2, and PER3). The various PER and CRY proteins can complex (dimerize) with themselves to form PER-PER homodimers or PER-CRY heterodimers. PER is phosphorylated by the kinase CK1 (Casein kinase 1 family of kinases) or other kinases, earmarking it for degradation. However, the PER-CK1 complex allows the CRYs to bind to form a CRY-PER-CK1 complex which prevents further phosphorylation and degradation of PER in the cytoplasm.

Within the complex of CRY-PER-CK1, CRY and PER are phosphorylated by other kinases which then allows the CRY-PER-CK1 complex to move into the nucleus and inhibit CLK-BMAL1 transcription of the *per* and *cry* genes forming the core negative limb of the TTFL. The CRY-PER-CK1 protein complex levels rise throughout the day, peak at dusk, and decline to their lowest level the following dawn. The stability/degradation rate of the CRY-PER-CK1 complex in the nucleus and the resumption of CLK-BMAL1 mediated transcription is a key process in setting the period of the clock.

It seems that CK1 and other kinases phosphorylate PER and target it for degradation, while at least one F-box protein (FBXL3) targets CRY proteins for degradation. The net result is that CRY and PER proteins fall to their lowest levels just before dawn. Light acts to upregulate *Per1* and *Per2* transcription and this allows the entrainment of the molecular clockwork to the dawn/dusk cycle. In many ways the CRY family of proteins remain enigmatic, not least because they play a key role in the negative component of the mammalian clock, yet form part of the light input pathway in the *Drosophila* clock.

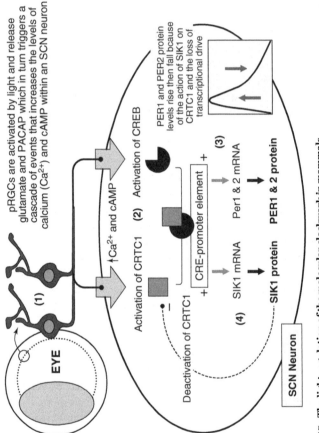

15. **The light regulation of the molecular clockwork in mammals.**

An interlocked secondary TTFL directs alternating activation and repression of BMAL1 expression (Figure 14). This occurs via the nuclear receptors RORα (RAR-related orphan receptor alpha) and REV-ERBα, respectively, via binding at ROR elements (retinoic acid-related orphan receptor response elements/ROREs) in the *Bmal1* promoter. Both *Rora* and *Rev-erba* have an E-box and are driven rhythmically via CLK-BMAL1 transcription. The rates of transcription and translation of these genes differ so that ROR peaks at dawn and REV-ERBα peaks at dusk, and this action on the *Bmal1* promoter ensures that BMAL1 levels rise at dusk, peak at dawn, and then fall throughout the day to their low point just before dusk. In this way BMAL1 levels cycle in antiphase to those of CRY and PER.

The *Dec1* and *Dec2* genes are mammalian homologs of the *Drosophila cwo* gene. DEC1 and DEC2 inhibit CLK-BMAL1 transcription and constitute the tertiary TTFL, which reinforces the action of CRY-PER-CK1 inhibition on CLK-BMAL1 transcription. Finally, and like *Drosophila*, the presence of an E-box in the promoter of downstream clock target genes gives rise to overt circadian rhythms in physiology and behaviour. However, it is also known that many clock-controlled genes do not possess an E-box. As a result, the nature of the circadian regulation in these genes remains uncertain.

The summary depicted in Figure 14 does not explain how the photosensitive retinal ganglion cells (pRGCs) (Chapter 4) bring about the entrainment of the molecular clockwork within SCN neurons. Circadian entrainment is surprisingly slow, taking several days to adjust to an advanced or delayed light/dark cycle. In most mammals, including jet lagged humans, behavioural shifts are limited to approximately one hour (one time zone) per day. At the molecular level, the effect of light on *per* gene induction is limited. After light exposure, *per* mRNA peaks after 1 hour, but then returns to baseline. Research at the University of Oxford has provided additional insights on how the molecular clock is entrained by light (Figure 15). Again, careful reading is needed!

The sequence of events that entrains the molecular clockwork of a SCN neuron to the solar day are summarized in Figure 15 and involve the following steps:

(1) Light is detected by the photosensitive retinal ganglion cells (pRGCs) within the eye. This induces the release of neurotransmitters (glutamate and pituitary adenylate cyclase-activating polypeptide/PCAP) from the pRGCs terminals which synapse with neurons in the ventral SCN. These neurotransmitters trigger a sequence of events that increase the levels of calcium (Ca^{2+}) and $3',5'$-cyclic adenosine monophosphate (cAMP) within an SCN neuron. Calcium levels rise as a result of influx from the extracellular medium or release from internal stores.

(2) Raised intracellular Ca^{2+} and cAMP activate two proteins: CREB-binding protein (CREB) and CREB-regulated transcription coactivator 1 (CRTC1), these work together and bind to a cAMP response element (CRE element) in the promoter of *Per1*, *Per2*, and *Slk1*.

(3) CRE activation of the *per* genes (+) leads to elevated *per* mRNA and increased levels of PER1 and PER2 protein. Changed levels of PER1 and PER2 act to shift the molecular clockwork, advancing the clock at dawn and delaying the clock at dusk. However, *per* mRNA and PER protein levels fall rapidly even if the animal remains exposed to light. As a result, the effects of light on the molecular clock are limited and entrainment is a gradual process requiring repeated shifting stimuli over multiple days. This phenomenon explains why we get jet lag: the clock cannot move immediately to a new dawn/dusk cycle because there is a 'brake' on the effects of light on the clock.

(4) The mechanism that provides this molecular brake is the production of SLK1 protein. SLK1 deactivates CRTC (–) by phosphorylation, so that it can no longer provide the co-transcriptional drive with CREB on the CRE promoter, and

78

transcription largely stops. This negative feedback turns off *Per1* and *Per2* transcription and translation, limiting the effects of light on the clock. *Slk1* mRNA and SLK1 protein levels also decline but more slowly than PER1 and PER2. The system then resets itself for possible light detection several hours later. Experiments on mice in which SLK1 has been suppressed show very rapid entrainment to simulated jet lag.

By limiting the shifting effects of light on the SCN, the circadian system of the animal is protected from abnormal light exposure at the wrong time of day. In addition, it may be important to buffer the effects of light on the SCN clock so that it is not pulled rapidly to a new phase, and in the process is uncoupled from the peripheral circadian network, resulting in internal desynchrony.

The depiction of the molecular clockwork of *Drosophila* and the mouse illustrated in Figures 14 and 15 is perhaps the most complete understanding to date of how genes and their protein products interact to give rise to behaviour. However, it is highly likely that additional genes and components of the core clock machinery have yet to be identified. Furthermore, the molecular story is even more complicated than depicted because there is a circadian rhythm to the structure of DNA itself, which in turn affects the regulation of clock genes.

Mammalian cells have about 2 metres of DNA. The packaging is accomplished with the aid of a class of proteins known as histones. DNA is wrapped around histones and the resulting macromolecule, which constitutes the chromosome fibre, is called chromatin. The rhythmic circadian gene transcription illustrated in Figures 14 and 15 is accompanied by a corresponding rhythmic modification of surrounding chromatin. It is not just the continual switching on and off of genes by the interplay of a host of molecules that results in a circadian rhythm, but chromatin itself changes shape in a daily fashion. The tighter the packaging of the local chromatin structure, the less probable it is that transcription and gene expression will occur. This shape-shifting which results from what are called histone acetylation and histone methylation accompanies both the activation and the repression of clock genes and clock-controlled genes.

Such mechanisms of histone acetylation and methylation are also at the heart of epigenetic modification. Epigenetics is a term used to describe cellular and physiological variations that result from external or environmental factors that modify gene expression, which may or may not be heritable. Linking circadian rhythms to epigenetics brings into play a host of intriguing possibilities. Not least that exposure to key environmental factors as we age could modify aspects of the circadian clockwork, which will in turn influence the pattern of sleep in old age. Perhaps the sleep we get in old age reflects what we did in our youth?

The discussion in this chapter has been limited to the molecular clocks of animals, and specifically *Drosophila* and mice. The elucidation of the molecular clockwork of fungi (e.g. *Neurospora crassa*), green plants (e.g. *Arabidopsis thaliana*), and cyanobacteria (e.g. *Synechococcus*) provides equally fascinating and complex stories with pioneering achievements by researchers including Jerry Feldman, Jay Dunlap, Steve Kay, Takao Kondo, Susan Golden, and Carl Johnson. In fungi, plants, and cyanobacteria the clock genes are all different from each other and different again from the animal clock genes, suggesting that clocks evolved independently in the great evolutionary lineages of life on earth. Despite these differences, all these clocks are based upon a fundamental TTFL. In Chapter 9 we will return to the topic of the molecular clockwork and consider another type of clock based upon feedback loops that do not involve a TTFL.

Chapter 6
Sleep: The most obvious 24-hour rhythm

The regular cycle of sleep and wakefulness is perhaps the most obvious 24-hour pattern of behaviour. The longest scientifically checked period without sleep in humans is 264.4 hours (11 days and 24 minutes). The 17-year-old student who went without sleep suffered serious cognitive and behavioural changes, problems with concentration and short-term memory, paranoia, and hallucinations. Most of us begin to fall apart after just one night without sleep, and after three nights of missed sleep we are functioning way below par.

We spend approximately 36 per cent of our entire lives asleep, and while asleep we do not eat, drink, or knowingly pass on our genes. This suggests that this aspect of our 24-hour behaviour provides us with something of huge value. If we are deprived of sleep, the sleep drive becomes so powerful that it can only be satisfied by sleep.

Many researchers assume that there must be a single overarching role for sleep embedded deep within our biology. Others reason that there is no single explanation and that different animals at different stages of their life cycle will use a period of sleep for different reasons, perhaps for energy conservation or to avoid predators. Yet others suggest that sleep is a trait that has no adaptive value but represents a by-product of some other truly adaptive trait yet to be discovered.

An alternative way to think about this issue is to break the problem into two related questions: (i) Why has almost all life evolved a 24-hour circadian pattern of activity and rest? (ii) What are the important processes that occur in the body during sleep?

Almost all life shows a 24-hour pattern of activity and rest, as we live on a planet that revolves once every 24 hours causing profound changes in light, temperature, and food availability. Diurnal and nocturnal species evolved numerous specializations that allowed optimum performance under the different conditions of light and dark. Life seems to have made an evolutionary 'decision' to be active at a specific part of the day/night cycle, and a species specialized to be active during the day will be far less effective at night. Conversely, nocturnal animals that are beautifully adapted to move around and hunt under dim or no light fail miserably during the day. The struggle for existence has forced species to become specialists and not generalists, and no species can operate with the same effectiveness across the 24-hour light/dark environment. Species are adapted to a particular temporal niche just as they are to a physical niche. Activity at the wrong time often means death.

Given that there are periods of activity and rest, what happens during those alternate states? Sleep may be the suspension of most physical activity, but a huge amount of essential physiology occurs during this time. Many diverse processes associated with the restoration and rebuilding of metabolic pathways are known to be up-regulated during sleep; genes associated with major metabolic pathways and the replenishment of neurotransmitter reserves in the nervous system are turned on during sleep; toxins that build up as the by-product of activity are processed and made safe and ready for removal during sleep; in humans and perhaps other animals with a complex brain, information received during the day is processed, and new memories and even new ideas are established. Indeed, 'sleeping on a problem' really can help the brain find new solutions to difficult issues.

During sleep the body performs a broad range of essential 'housekeeping' functions without which performance and health during the active phase deteriorates rapidly. But these housekeeping functions would not be why sleep evolved in the first place. Such essential activities are demanded by our biology and need to occur at some point over the rest/activity cycle. Evolution has allocated these key activities to the most appropriate time of day. So memory consolidation occurs after activity, during sleep, when the brain is not being swamped with new sensory information and has the capacity and available energy to perform the task optimally. In the same way, toxin clearance and the rebuilding of metabolic pathways needs to occur after the toxins have built up and energy substrates have been depleted during activity.

We do not know why humans sleep on average 8 hours a day, or why some species sleep for 19 hours and others for only 2 hours, but this undoubtedly arises from a complex set of competing factors and interactions. To survive and thrive, individuals need to balance the requirements for food, water, and breeding partners with the problems of encountering predators, pathogens, and physical danger. Once a stable rest/activity pattern has evolved for a given species, then essential housekeeping processes will be incorporated into this temporal structure at an appropriate phase. Such housekeeping activities will then act to reinforce sleep behaviour. The integration of multiple and often unrelated elements of physiology into the rest phase of the 24-hour activity/rest cycle represents an evolutionary leap forward, transforming mere inactivity and the avoidance of harm into the complex sleep behaviour we see in ourselves and many other animals.

In short, sleep has probably evolved as a species-specific response to a 24-hour world in which light, temperature, and food availability change dramatically. Sleep is a period of physical inactivity when individuals avoid movement within an environment to which they are poorly adapted, while using this time to undertake essential housekeeping functions demanded by their biology.

In 1982, Alexander Borbély, at the University of Zurich, proposed the 'two-process model' of sleep regulation which provides a conceptual framework for understanding the timing and structure of sleep/wake behaviour. The two-process model describes interacting drivers involving a homeostatic process (S), which increases as a function of the duration of wakefulness, and a circadian process (C), which determines the timing of sleep and wakefulness (Figure 16).

From the moment an individual wakes (Figure 16) the sleep pressure (S) builds throughout the period of wake then dissipates during sleep. The circadian system (C) 'time-stamps' the neural circuitry for sleep/wake by defining when it is biologically appropriate to sleep or be awake. As sleep pressure builds during the day it is opposed by the circadian drive for wakefulness. But as night approaches, the circadian drive for wake declines. This circadian decline in wakefulness (which can also be considered as a drive for sleep), combined with a high degree of sleep pressure, initiates sleep. This interaction between Process C and Process S has been called the 'sleep gate'. Sleep pressure within the sleep gate will be highest during the first part of the night but is increasingly reduced as the homeostatic drive for sleep dissipates. This, combined with the increased circadian drive for wakefulness towards dawn, will drive an individual to wake up. Of course, the human sleep/wake cycle is heavily influenced by social factors, such as the alarm clock signalling the time to get up. But when these social cues are removed there is a clear sleep/wake cycle dictated by the balance between Process S and Process C.

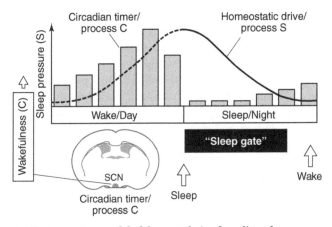

16. The two-process model of sleep regulation for a diurnal mammal. A 24-hour circadian signal (C, grey bars) arising from the suprachiasmatic nucleus (SCN) and a homeostatic driver (S, dotted line) interact to determine the timing, duration, and structure of sleep.

The relationship between Process S and Process C was demonstrated in experiments with rats with lesions that destroyed their SCN. Lacking any circadian drive, these rats would cycle between relatively short bouts of sleep and rest driven by a build-up and then dissipation of sleep pressure. Critically, when rats were kept awake by gentle handling they would sleep longer during the next sleep episode, demonstrating that the duration of wake would influence the duration of subsequent sleep—sleep was being homeostatically regulated.

Several chemical mediators have been implicated in driving sleep homeostasis. Such a chemical would be expected to accumulate after prolonged wakefulness and decline during sleep. Adenosine has emerged as a strong candidate. Adenosine forms from the breakdown of adenosine triphosphate (ATP), which is the primary energy currency in cells. As a result adenosine build-up provides a good marker of activity. Adenosine increases in the brain during waking hours and after sleep deprivation. Furthermore, perfusion of adenosine into the brain of freely moving rats reduces wakefulness and activates neurons associated with sleep promotion, and these responses are lost in mice lacking the adenosine A_2 receptor. Caffeine is a potent stimulant that can certainly reduce sleepiness in humans and functions by blocking these adenosine A_2 receptors. In addition to adenosine, there are other factors including prostaglandin which has been proposed as acting along with adenosine to provide the homeostatic drive for sleep.

The two-process model has been very powerful in understanding the basic interactions between the circadian system and homeostatic drivers regulating sleep (Figure 16), but in reality the regulation of sleep is likely to be much more complicated. Not least because in humans and other animals sleep can be 'biphasic' or even 'polyphasic', with two or more periods of sleep separated by short periods of wake. How such fragmentary sleep is generated is uncertain and will require additional inputs to the model depicted in Figure 16.

Beyond the circadian and homeostatic drivers of sleep, the sleep/wake cycle involves a highly complex set of interactions between multiple neural circuits, neurotransmitters, and hormones, none of which are exclusive to the generation of sleep. These systems combine to change sleep/wake states in a way which is analogous to having a 'flip flop' switch common in electronic circuits. Transitions occur over seconds to minutes (depending upon the species being studied), and result in clear-cut changes in behavioural states. Intermediate states would render an animal at a distinct selective disadvantage. The major brain structures and neurotransmitter systems involved in the sleep/wake cycle are summarized in Figure 17.

During wake, orexin neurons in the lateral hypothalamus project to and excite (+) different populations of monoaminergic neurons across the hind- and mid-brain which release histamine,

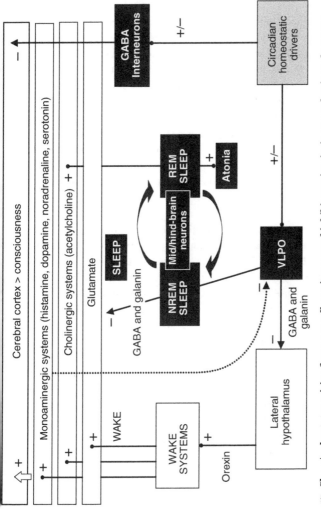

17. **Sleep/wake states arising from mutually excitatory and inhibitory circuits in a way that is analogous to having a 'flip flop' switch common in electronic circuits.**

dopamine, noradrenaline, and serotonin; cholinergic neurons in the hind-brain release acetylcholine; and an important group of neurons release glutamate (Figure 17). Collectively these neurotransmitters drive wakefulness and consciousness within the cortex. The monoaminergic neurons project to (dotted line in Figure 17) and inhibit (–), the ventrolateral preoptic nuclei (VLPO). During sleep, circadian and homeostatic sleep drivers activate the VLPO which releases gamma-aminobutyric acid (GABA) and galanin to inhibit the orexin neurons in the lateral hypothalamus and the aminergic, cholinergic, and glutamatergic neuronal populations (–). Further, a subpopulation of interneurons in the cortex project long distances to the cerebral cortex and release the inhibitory neurotransmitter GABA during sleep. The activation of these GABA neurons is proportional to homeostatic drive for sleep.

Sleep in mammals is characterized as either rapid eye movement (REM) or non-rapid eye movement (NREM) states. The NREM–REM switch occurs every ~70–90 minutes and is driven by a network of neurons within the mid- and hind-brain (see Figure 17). During REM sleep aminergic neurons remain inhibited, but cholinergic neurons are activated (+). REM-on neurons project to the spinal cord and drive muscle paralysis (atonia).

In relation to sleep, melatonin receptors located on SCN neurons are thought to detect nocturnal melatonin and provide additional feedback for clock entrainment, reinforcing light entrainment signals from the eye. However, the direct action of melatonin in sleep is not very clear even though melatonin is commonly, though wrongly, called the 'sleep hormone'.

Melatonin is synthesized mainly in the pineal gland, although the retina and other regions of the body can also produce small amounts. The pineal is tightly regulated by the SCN to produce a circadian pattern of melatonin release, with levels rising at dusk, peaking in the blood around 02.00–03.00 a.m. and then declining before dawn. Light, detected by the eyes, also acts to inhibit melatonin production acutely. As a result, melatonin acts as a biological marker of the dark and the duration of melatonin release in mammals provides a critical signal for day length and the regulation of seasonal physiology (Chapter 8).

While melatonin production occurs during sleep in diurnal mammals, nocturnal species such as rats and hamsters also produce melatonin at night when they are active. Sleep propensity in humans is closely correlated with the melatonin profile but this may be correlation and not causation. Indeed, individuals who do not produce melatonin (e.g. tetraplegic individuals, people on beta-blockers, or pinealectomized patients) still exhibit circadian sleep/wake rhythms with only very minor detectable changes.

Another correlation between melatonin and sleep relates to levels of alertness. When melatonin is suppressed by light at night alertness levels increase, suggesting that melatonin and sleep propensity are directly connected. However, increases in alertness occur before a significant drop in blood melatonin. Furthermore, increased light during the day will also improve alertness when melatonin levels are already low. These findings suggest that melatonin is not a direct mediator of alertness and hence sleepiness.

Taking synthetic melatonin or synthetic analogues of melatonin produces a mild sleepiness in about 70 per cent of people, especially when no natural melatonin is being released. The mechanism whereby melatonin produces mild sedation remains unclear. However, this ability to reduce the time it take to fall asleep, and melatonin's ability to help entrain the circadian clock, makes melatonin a moderately useful agent in treating circadian rhythm disruption such as jet lag.

In our 24/7 society there are a multitude of other factors that impact upon sleep. These include: children, pets, noise and light pollution, exposure to hot or cold, intractable pain, a snoring bed partner, chronotype, work demands, environmental temperature, medications, when we eat, exercise, and social media. Many of these factors can be relatively simple to deal with at an individual level—using black-out curtains or wearing an eye mask, using earplugs, wearing bedsocks, taking the TV, laptop, smartphone, and pets out of the bedroom, drinking less alcohol, exercising earlier, and going to bed earlier.

But there are areas which are more difficult to handle. The impact of children on parents' sleep is life-changing. It can take several months before babies become fully adapted to the 24-hour light/dark cycle and several years for children's sleep patterns to resemble those of adults. A shift from living in extended family groups where childcare was shared, to a more nuclear family structure, has placed additional burdens upon parents and invariably the mother. Several genes have been identified which are associated with sleep timing and sleep duration. If a bed partner is a short or long sleeper, or wakes early or goes to bed late, then this will have a major influence upon a partner's sleep—and have knock-on consequences upon the relationship. Chronotype also influences the tendency to abuse drugs and alcohol (Chapter 3). There are no easy solutions, but an awareness of the importance of sleep, and prioritizing sleep whenever possible, will improve the quality of life across all sectors of society.

Chapter 7
Circadian rhythms and metabolism

Life is a continual battle to stay in a non-equilibrium energy state. Living things must obtain energy from the environment and effectively use it to drive metabolic reactions. Plants capture energy from the photons of sunlight; bacteria in the Sirena Deep of the Mariana Trench feed off the products of chemical reactions between rock and seawater; we eat other animals, and plants and fungi. The constant energy cycle is the fundamental fact of life because when it stops we are at energy equilibrium, or in other words dead!

Feeding behaviour in animals is modified by food availability, satiation, hunger, social factors, and of course circadian timing. The circadian system plays a fundamental role in coordinating anabolic (constructing molecules from smaller units) and catabolic (breaking down large molecules into smaller parts) metabolic pathways. Energy intake and expenditure in animals varies according to sleep/wake and fasting/feeding periods, the timing of which depends on whether the species is diurnal or nocturnal. During the active period, eating ensures the uptake of energy substrates such as carbohydrates, lipids, and amino acids. In mammals, these are either stored or processed immediately. Stored energy substrates such as glycogen and fat in the liver are metabolized into glucose to maintain appropriate basal energy expenditure both during activity and rest. The basic regulation of blood glucose is summarized in Figure 18.

The key interactions in glucose metabolism are summarized in Figure 18. Feeding changes blood glucose. In response to low blood glucose (hypoglycaemia) the alpha cells of the pancreas are stimulated to release glucagon. Glucagon stimulates the liver to break down glycogen to be released into the blood as glucose and activates gluconeogenesis, which is the conversion of proteins (amino acids) and fats (triglycerides) into glucose. Glucagon also stimulates the breakdown of stored fat (triglycerides) in adipose tissue into free fatty acids which are then converted in the liver into glucose. The net effect is to raise blood glucose. By contrast, in response to high blood glucose (hyperglycaemia), the beta cells of the pancreas release insulin. Insulin regulates how the body uses and stores glucose and fat. Insulin enhances entry of glucose into metabolically active cells such as muscle; it stimulates the liver to convert glucose into glycogen for storage; and in adipose tissue it suppresses the breakdown of fat into free fatty acids and promotes the synthesis of stored fat from free fatty acids. The net effect is to lower blood glucose.

The circadian timing of energy metabolism (Figure 18), of which blood glucose levels are a major part, is a multi-layered system, involving control by the SCN as well as by peripheral clocks in organs such as the liver, pancreas, muscle, and white adipose tissue. The SCN regulates all aspects of glucose homeostasis, including glucose production, glucose uptake, and insulin release and insulin sensitivity. Insulin plays a key role by inhibiting glucose production in the liver and by stimulating glucose uptake by insulin-sensitive tissues (mainly skeletal muscle and adipose tissue). But it is a complicated process. For example, human skeletal muscle consumes more glucose in the morning than in the evening, and this rhythm is the net result of rhythms in insulin production, insulin sensitivity, and insulin-independent glucose uptake.

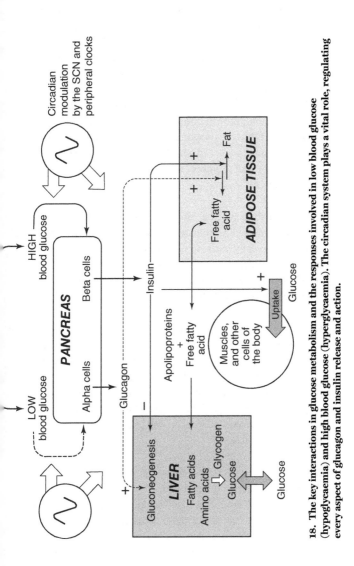

18. The key interactions in glucose metabolism and the responses involved in low blood glucose (hypoglycaemia) and high blood glucose (hyperglycaemia). The circadian system plays a vital role, regulating every aspect of glucagon and insulin release and action.

In addition to the mechanisms described in Figure 18, there are several hormones that play an important role in glucose regulation: epinephrine/adrenalin from the adrenal medulla enhances gluconeogenesis and antagonizes the action of insulin; adrenocorticotropic hormone (ACTH) from the anterior pituitary stimulates the release of cortisol/glucocorticoids from the adrenal cortex which enhances gluconeogenesis and antagonizes the action of insulin; growth hormone released at night from the anterior pituitary antagonizes the action of insulin; and thyroxine from the thyroid gland enhances the release of glucose from glycogen and promotes the absorption of sugars from the intestine.

Diabetes mellitus type I is caused by insufficient or non-existent production of insulin, while type II is primarily due to a decreased response to insulin in the tissues of the body (insulin resistance). Both types of diabetes result in too much glucose remaining in the blood (hyperglycaemia). This is referred to as impaired glucose tolerance (raised blood glucose levels).

Glucose is the near universal fuel of living things, and maintaining an appropriate level of blood glucose across the 24-hour day is essential for normal physiology. This is particularly so for the central nervous system, which can neither synthesize nor store glucose for its normal cellular function. The human brain may be only about 2 per cent of the body mass but uses about 20 per cent of our energy intake in the form of glucose and it is active by night as well as day.

All the metabolic pathways regulating blood glucose are either directly or indirectly under circadian control, modulated by the SCN and/or the peripheral clocks within the key metabolic organs and tissues. The liver and adipose tissue are innervated indirectly by the SCN via extensive sympathetic as well as parasympathetic projections. Furthermore, digestive processes are all under circadian control. In diurnal animals such as humans, there is a circadian rhythm in saliva production that rises over the day and falls at night; the stomach empties faster after identical meals in the morning than in the evening; colonic motility has a circadian rhythm, with movement during the day and little during the night; while gastric acid secretion has a daily rhythm with increased production in the evening anticipating prior food intake.

Recent studies have highlighted the importance of meal timing on the circadian network. For example, if nocturnal mice are only allowed access to a high-fat diet during the light phase of a light/dark cycle then metabolism can be severely disrupted. Mice get fat and show impaired glucose tolerance (raised blood glucose levels) compared to night-fed mice. Food restriction has also highlighted another important finding. Allowing mice or rats access to food only in the middle of the day changes the phase of clock gene expression in metabolically active organs such as the liver, heart, skeletal muscle, and adipose tissue to the time of restricted feeding. Significantly, the phase of clock gene expression in SCN cells remains unaffected. The net result of food intake at an inappropriate environmental time is the phenomenon of 'internal desynchrony', which refers to the misalignment between central and peripheral clocks and between the clocks of different peripheral organs such as the liver, adipose tissue, and muscles.

The timing of food intake, or 'chrononutrition', has emerged as a new field of research which incorporates the idea that the time of food intake, along with the amount and content of food, is critical for the overall health of the individual. For example, glucose tolerance decreases from the morning to the evening in healthy individuals. This means that the glucose is utilized faster in the morning compared to the evening and at night. In healthy individuals, given the same meal in the morning at 08.00 and then again 12 hours later at 20.00, blood glucose levels were shown to be significantly higher (17 per cent) after the evening meal versus the morning meal—the subjects showed lower glucose

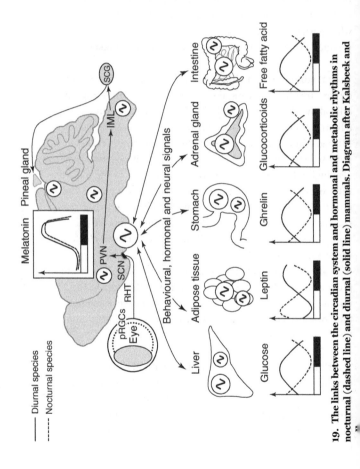

19. The links between the circadian system and hormonal and metabolic rhythms in nocturnal (dashed line) and diurnal (solid line) mammals. Diagram after Kalsbeek and

tolerance in the evening. The researchers then simulated a night shift work pattern, where participants were scheduled to sleep during the day. After only three days of this circadian rhythm disruption glucose tolerance was even worse in the evening and subjects showed signs of insulin resistance (decreased response to insulin). The authors concluded that circadian misalignment impairs glucose tolerance and that this provides a mechanism to help explain the increased risk of diabetes in shift workers.

The continuous homeostatic fine-tuning of blood glucose levels in response to environmental changes, such as a meal, occurs within a circadian framework that anticipates daily energy needs. During activity (the feeding period) blood glucose is mainly of dietary origin, while during sleep (starvation) glucose is progressively released from endogenous glucose production from gluconeogenesis in the liver (see Figure 18). The net result is a large daily fluctuation in the glycogen content of the liver as it acts to sustain blood glucose levels across the differing demands of the 24-hour rest/activity cycle.

The hypothalamus sits at the centre of physiological control, regulating metabolic processes that include feeding behaviours, energy intake/expenditure, glucose and lipid metabolism, and thermoregulation, and all of these systems are regulated by the circadian system (Figure 19).

The photosensitive retinal ganglion cells (pRGCs) detect light/dark information and via the retinohypothalamic tract (RHT) convey this information to the SCN. The SCN then transmits this temporal information to peripheral clocks in the brain and the rest of the body via behavioural, hormonal, and neural signals. Peripheral signals, in turn, can feed back to regulate the SCN and other brain clocks. The central role of the SCN has been shown by lesion studies which abolished daily rhythms in plasma glucose, insulin, stored glycogen, and other key metabolic substrates. The SCN connects to other key nuclei within the hypothalamus, including the arcuate nucleus (ARC), ventromedial hypothalamus (VMH), lateral hypothalamus (LH), dorsomedial hypothalamus (DMH), and paraventricular nucleus (PVN). Reciprocal connections between the ARC nucleus and the SCN have been shown to be especially important, forming the basis by which the SCN is informed about metabolic status and the ARC about the time of the day. Sensors within the ARC nucleus detect nutrients such as glucose and lipids, and circulating metabolic hormones such as leptin, insulin, thyroid hormone, and ghrelin released from glands and tissues in the intestine, pancreas, stomach, and adipose tissue. In diurnal and nocturnal species, before waking glucose production and glucose concentrations are increased while glucose utilization is high. Thus glucose production and utilization both increase at the beginning of the activity period and show a clear daily rhythmicity.

The SCN regulates the nocturnal release of melatonin in nocturnal and diurnal species by conveying photic information to the PVN, and from the PVN autonomic fibres project to the intermediolateral column (IML) of the spinal cord and then the superior cervical ganglia to reach the pineal gland (Figure 19). Melatonin seems to play a role in rodent metabolism as pinealectomized rodents show disturbed 24-hour rhythms of plasma glucose concentration, loss of the daily rhythm of glucose-induced insulin secretion, impaired glucose tolerance, and decreased adipose cell responsiveness to insulin.

The liver is the main source of stored glucose, and via the autonomic nervous system the liver is connected to the SCN. If this SCN to autonomic nervous system to liver connection is blocked or disrupted the daily rhythm in glucose metabolism is lost, again demonstrating the key role of the SCN in glucose metabolism. Liver clock cells continue to 'tick', but without the SCN they all drift out of phase and the coordinated rhythm is lost. The SCN does not project directly to the autonomic nervous system, rather it projects to the PVN which then regulates autonomic signalling to peripheral organs via its extensive projections to sympathetic and parasympathetic neurons in the spinal cord and in the brainstem. Stimulating the activity of PVN neurons results in hyperglycaemia (high blood sugar) by activating sympathetic input to the liver to promote gluconeogenesis.

Another target area of the SCN is the LH. The LH contains the main population of orexin-containing neurons which are not only involved in arousal and sleep (Chapter 6), but also play a key role in food intake and energy metabolism. Orexin neurons show a marked day/night rhythm, with peak activity during the waking period in both nocturnal and diurnal species. Increased activity of the orexin neurons at the end of the sleep period, again influenced by the SCN, not only results in arousal but also in gluconeogenesis by the liver. Increased glucose levels in turn allows an increased heart rate, body temperature, and glucose uptake by the muscles.

Hormonal signals from the periphery provide information to the brain about energy demands. Leptin secreted by the white adipose tissue acts on multiple targets but notably the ARC, which integrates and responds to satiety and hunger signals and forms the origins of the central neural response to perturbations in energy balance. Neuropeptide Y (NPY)-producing neurons in the arcuate nucleus stimulate food intake, whereas arcuate nucleus neurons that release the proopiomelanocortin (POMC)-derived peptide, alpha-melanocyte-stimulating hormone (alpha-MSH), reduce food intake. Leptin inhibits NPY neurons and activates POMC neurons, thus resulting in reduced food intake.

There is a 24-hour rhythm of leptin release which is under the control of the SCN via its autonomic input to the adipose tissue. The anorectic (reduced food intake) action of leptin is opposed by ghrelin which is primarily secreted by the stomach. Ghrelin stimulates feeding by activation of neurons in the ARC, resulting in increased release of NPY in the PVN. Plasma levels of ghrelin change with the feeding cycle but are also under circadian control.

In nocturnal rodents, plasma levels of ghrelin increase during the rest/sleep period (light) in anticipation of food intake. Ghrelin also feeds back to the SCN, altering clock gene expression and providing an endocrine signal that allows the stomach to communicate with the central clock.

The PVN acts as an important integrator for metabolic homeostasis. It receives neural and humoral signals from the SCN while neurons containing corticotrophin-releasing hormone stimulate the rhythmic secretion of adrenocorticotropic hormone (ACTH) from the anterior pituitary gland which then stimulates the production and release of glucocorticoids (GCs) from the cortex of the adrenal gland.

GCs play a vital role in the regulation of energy metabolism. Too much leads to hyperglycaemia (excess blood sugar), hypertension (high blood pressure), sleep disturbance, body weight gain, and other metabolic abnormalities. GCs are also critical for glucose homeostasis by stimulating liver gluconeogenesis. GCs may also directly regulate clock gene expression in metabolically active tissues, such as the liver and kidney. In nocturnal rodents, the peak of plasma GC is phase-locked to activity onset. In the context of sleep and circadian rhythm disruption (SCRD), stress can markedly alter levels of GC and mask the endogenous rhythm of GC. Light in nocturnal species causes the acute release of GCs via activation of the sympathetic nervous system, and this action is independent of ACTH release or the SCN.

Lipid metabolism, like glucose, also shows a robust daily rhythm aligned to providing the appropriate energy requirements across the sleep/wake and fasting/feeding cycles. Lipid biosynthesis, transport, and breakdown for energy are under circadian control and tightly coupled. Lipid molecules, including triglycerides and cholesterol, cannot circulate easily in the blood and are transported by apolipoproteins. In nocturnal rats and mice the nocturnal rise in plasma triglycerides and cholesterol is caused by circadian changes in apolipoproteins. The gastrointestinal tract absorbs lipids and absorption is higher during the active period and lower during the resting period. Intestinal epithelial cells show rhythms in clock gene expression that are synchronized by both SCN signals and the availability of food. Clock gene expression within the intestinal epithelial cells appears to control the daily expression of proteins involved in lipid absorption, including apolipoproteins and nocturnin.

Again, as with glucose regulation, there is good evidence that the SCN controls the day/night rhythms in lipid metabolism via the autonomic nervous system. Adipose tissue is innervated extensively by sympathetic as well as parasympathetic projections from the SCN. Activation of the sympathetic branch stimulates catabolism—increased lipolysis—and the hydrolysis of triglycerides into glycerol and free fatty acids, while parasympathetic activation leads to anabolic activity and an increase in the insulin-mediated uptake of glucose and free fatty acids in adipose tissue.

The complex interactions between the SCN, hypothalamus, hormones, and a host of other signalling molecules involved in the circadian regulation of metabolism is mirrored at the subcellular level. Gene expression profiles have been measured across the circadian cycle in the mammalian liver, skeletal muscle, pancreas, heart, and brown and white adipose tissue. The numbers of genes showing 24-hour rhythmic changes ranges from about 5 per cent up to 20 per cent depending on the tissue. Different tissues show only a limited overlap in rhythmic genes, suggesting that expression is regulated in a tissue/organ-specific manner. Among the rhythmic genes identified, many have roles in biosynthetic and metabolic processes, including cholesterol and lipid metabolism, glycolysis and gluconeogenesis, oxidative phosphorylation, and detoxification pathways.

There is a close association between the molecular clockwork of metabolic cells and the regulation of their metabolic activities, including glucose and lipid metabolism and the synthesis of cholesterol and bile acids. The core clock oscillator constitutes a transcription–translation feedback loop (TTFL) involving BMAL1, CLOCK, PERs, and CRYs and multiple other transcription factors such as DBP, DEC2, HLF, and TEF (Figure 14). The transcriptional control of metabolism by the molecular clockwork is illustrated in Figure 20, and can be summarized as follows. Peripheral clocks are regulated both by the SCN and metabolic signals including hormonal signals (e.g. insulin) and feeding/fasting behaviours that change the levels of glucose. Within the cells of key metabolic organs (e.g. the liver, adipose tissue, stomach, adrenal glands, and intestine) these environmental signals alter the phase of the core molecular clockwork and key outputs regulators from the clock. These proteins then act to regulate their metabolic target genes. For example, DBP (D site of albumin promoter (albumin D-box) binding protein) binds to an upstream promoter to regulate the insulin gene in pancreatic beta cells; HLF (Hepatic leukaemia factor), regulates aspects of liver function; and TEF (thyrotroph embryonic factor) is involved in thyroid-stimulating hormone release.

The circadian coordination of metabolism also involves members of the reverse-erb (REV-ERB) receptor family, retinoic acid orphan receptors (ROR), PPARs (peroxisome proliferator-activated receptors) and other nuclear receptors. Metabolic regulators, such as REV-ERBα and ROR, also participate directly in the clock mechanism by regulating BMAL1 transcription. In addition, hepatic PPARα, which is activated by fatty acids, is regulated rhythmically by CLOCK and BMAL1 and is also regulated by glucocorticoids. These transcriptional regulators in turn interact with metabolic target genes associated with glucose, cholesterol, and triglyceride metabolism. Such target genes in the liver include: glycogen synthase, involved in converting glucose to glycogen; HMG-CoA reductase, which is the rate-controlling enzyme that produces cholesterol; CYP7A1, which is a rate-limiting

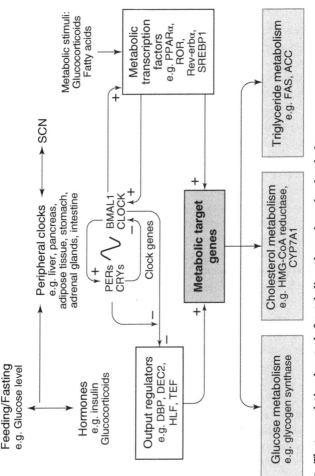

20. The transcriptional control of metabolic pathways by molecular clocks.

enzyme in bile acid synthesis; and Acetyl-CoA carboxylase and fatty acid synthase, which are involved in catalysing the synthesis of fatty acids.

This regulation can be immensely complex, with multiple interlocking feedback loops between the clock and metabolic genes/proteins. For example, transcriptional regulation of rhythmic CYP7A1 in the liver is driven by DBP, the clock protein DEC2, and by nuclear receptors including PPARα. In both the liver and adipose cells PPARα also regulates REV-ERBα expression, while ROR and REV-ERBα regulate lipid metabolism as well as being involved in CLOCK and BMAL1 expression!

There is a continual interaction, in which circadian clocks within the major metabolic organs coordinate metabolic timing, and then metabolic products and activated proteins/signalling molecules feedback to adjust the timing of their molecular clockwork. For example, studies have shown that the protein deacetylase, Sirtuin 1 (SIRT1), controls both glucose and lipid metabolism in the liver, promotes fat mobilization, controls insulin secretion in the pancreas, senses nutrient availability in the hypothalamus, and influences obesity-induced changes in metabolism. In addition, SIRT1 regulates CLOCK/BMAL1 function and so exerts a direct regulatory input on PER/CRY expression. SIRT1 also promotes the degradation of PER, while CLOCK and BMAL1 are required for the circadian expression of SIRT1 in the liver. In mice, the lowering of CLOCK or BMAL1 levels by genetic manipulation increases liver insulin resistance—the failure of the liver to respond to insulin, leading to high blood sugar. It seems that CLOCK/BMAL1 binds to the SIRT1 promoter to enhance its expression and regulate hepatic insulin sensitivity by SIRT1. Mice with circadian rhythm disruption show decreases in liver BMAL1 and SIRT1 levels and this is correlated with increased liver insulin resistance.

When mice are given access to a high-fat diet across the 24-hour day they feed much more during the sleep (light) phase than normal. They also show blunted oscillations of REV-ERBα expression in the liver, as well as increased fat deposits and signs of liver disease. One explanation is that the SCN is entrained by light and sends out a signal promoting sleep and the mobilization of lipids for energy use while another zeitgeber, food, is instructing the cell that it is not sleep time but activity time and fat should be stored for use later during sleep. This internal desynchrony, and mixed circadian signalling between central and peripheral clocks and inappropriately timed metabolic feedbacks to the circadian system, has important implications for health. It seems likely that the health problems seen in shift workers and in mice fed during their sleep phase could arise from an uncoupling of the SCN and the peripheral clocks due to conflicting signals. Eating at an inappropriate time of the day means that the peripheral clocks in the liver, adipose tissue, pancreas, and muscle are at an entirely different phase from the SCN, and the molecular clocks are further 'confused' by inappropriately timed levels of glucose, insulin, glucocorticoids, output regulators, and metabolic transcription factors. In short, circadian meltdown.

In the past decade, multiple cellular and subcellular mechanisms linking the circadian and metabolic systems have been discovered, allowing for the first time a mechanistic understanding of how circadian rhythm disruption can lead to metabolic abnormalities and vice versa. This emerging understanding has been driven in large part because the tools necessary to investigate circadian and metabolic connections have only become available in recent years. Methodologies that include microarrays, fluorescent reporters, transgenic organisms, RNA interference, and industrial-scale gene expression assays have transformed our understanding, and represent a spectacular example of how genuinely new and profound knowledge, with important clinical implications, can arise by using such advanced biotechnologies.

Chapter 8
Seasons of life

Some 4.5 billion years ago, the nascent and still condensing
earth was in an almighty collision with another proto planet.
The bang ejected a moon-sized chunk and our planet's spin axis
was tilted with respect to the plane of its orbit about the sun to
an angle of 23.5°. That is one explanation for the tilt. There
are others.

Whatever happened all those billions of years ago, the net result is
that we now have seasons (see Figure 21). Since the earth's axis is
tilted by 23.5°, different parts of the earth are oriented towards
the sun at different times of the year. Summer is warmer than
winter (in each hemisphere) because the sun's rays hit the earth at
a more direct angle during summer and the days are much longer
than the nights. During winter the sun's rays are more oblique and
the days are very short. At the equator, the earth's surface receives
close to 12 hours of light per day (sunrise to sunset) all year long.
As one proceeds north or south of the equator, the annual
variation in day length becomes progressively more extreme,
increasing from 0 hours right at the equator to 24 hours in the
summer at latitudes above 67°. Up to 30° latitude either side of
the equator, the daily temperature is above 15°C for ten months of
the year, while at 48° latitude the 15°C isotherm lasts only three
months. With each season there are changes in day length, wind
speed and direction, temperature, and rainfall. These changes

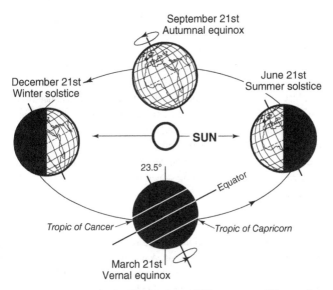

September 21st
Autumnal equinox

June 21st
Summer solstice

December 21st
Winter solstice

SUN

23.5°

Equator

Tropic of Cancer

Tropic of Capricorn

March 21st
Vernal equinox

21. The Earth's axis is tilted by 23.5°, and different parts of the earth are oriented towards the sun at different times of the year. This creates the seasons.

dictate the optimal times for plants and animals to initiate growth, development, and reproduction.

Plants are sessile and cannot escape or hide from seasonal change. Getting the timing right so that germination, bud burst, flowering, dormancy, and a host of other critical processes are aligned with seasonal changes is essential for survival. Animals have more choices. They can migrate, hibernate, or, like plants, adapt their physiology and behaviour to cope with the changes.

Survival in any individual is about anticipating daily and seasonal events and timing critical events in the life cycle to optimize reproductive success. The changing environmental conditions include not just features such as day length, rainfall, and temperature, but the entire ecological web. The period during which favourable environmental conditions occur, to a large extent determines the timing of growth, reproduction, and survival of offspring. These critical events in organisms at the bottom of the food chain (e.g. algae and plants) dictate the timing of life cycle events in organisms higher up the food chain. At its simplest, many animals produce offspring in spring because this is when food is at its most abundant, and there is sufficient time for the young to develop and survive the harsh conditions of their first winter. To do this, the parents have to perfect the timing of mating, conception, and birth. Failure often means death, and so there has been a very strong selection pressure to get the timing right.

In theory any regular environmental time signal could be used to time seasonal reproduction. Temperature shows a seasonal variation, but it is not very precise. A warm January or February can be followed by a cold snap in March. Evolution will favour those signals that are the most reliable from year to year, and by far the most stable indicator of the time of year is photoperiod—or day length. The annual cycle of variation in photoperiod is consistent from year to year. In the northern hemisphere the summer solstice is always around 20/21 June and the winter solstice around 21/22 December. A breathtaking diversity of life, from unicellular organisms to giant redwood trees and the blue whale, use day length to anticipate and prepare for the changing seasons.

In the 1920s, Wightman Garner and Henry Allard discovered photoperiodism in plants showing that changes in day length, in the absence of any other clues, trigger flowering. Many plants use what is effectively a photoperiodic calendar to 'know' the time of year. For very many plants in the non-equatorial regions there is a species-specific critical day length, the point at which the photoperiod (day length versus night length) switches from being non-inductive to inductive. Plants such as the soya bean, which flower in the autumn, are called short-day plants (SDPs), because they are triggered to flower when the day length is shortening.

Other plants, such as barley, flower in the spring/summer when the days are lengthening, and these are named long-day plants (LDPs). Some plants, such as rice, corn, and the cucumber, are not sensitive to the photoperiod and are called day-neutral plants.

Over the past 140 million or more years, the flowering angiosperms have developed a set of light-sensitive receptors and a signalling system linked to a molecular clock within the plant that enables them to define both the time of day and the time of year. Many plants, for instance, remain vegetative before winter, avoiding frost damage to their developing reproductive organs, but then flower rapidly in spring, to allow flowering and seed production before the onset of heat and water restriction in summer.

In the 1930s, Erwin Bünning proposed a mechanism based on his work on plants and insects. He reasoned that the daily rhythm had two alternating phases of about 12 hours each, which he distinguished as a light-loving (photophilic) period during the day and a dark-loving (scotophilic) period in the dark. Light falling on a plant at a specific phase during the photophilic phase will enhance flowering initiation, and during the scotophilic period will inhibit it. There is a specific, critical, photoinducible phase, and when this interacts with light, seasonal events are triggered. This process is summarized for an LDP in Figure 22, and illustrates Bünning's hypothesis or the 'external coincidence model' of photoperiodic regulation. This model proposes that a circadian clock is entrained to the light/dark cycle, which in turn regulates the rhythm of expression of a molecule (molecule 'X') that can stimulate (acting as a 'florigen') or inhibit flowering. The rhythm in molecule X is entrained such that it rises in anticipation of dusk. In an LDP, if light falls upon molecule X it will trigger flowering (photoinduction). During long nights (autumn/winter) the photoinducible phase (● in Figure 22) is not exposed to light, but during short nights (spring/summer) the photoinducible phase is exposed to light and a photoperiodic response is triggered. In the LDP shown in Figure 22, exposure of the

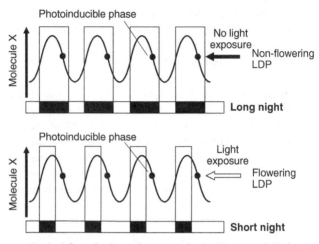

22. **Bünning's hypothesis, or the 'external coincidence model' of photoperiodic regulation.**

photoinducible phase to light promotes flowering, but in the case of an SDP light will inhibit flowering.

Arabidopsis thalania is a long-day flowering plant and is the 'workhorse' in plant circadian studies. Experimental proof for the involvement of a circadian clock in photoperiodic flowering came from *Arabidopsis* mutants in which the same mutation disrupted both daily and photoperiodic timekeeping. However, the precise mechanisms have turned out to be far more complex than Bünning anticipated, involving an intricate interaction between the molecular clockwork of the plant combined with light-signalling molecules located in the leaves. There is indeed 'external coincidence' (see Figure 22), or the existence of what is effectively a circadian rhythm of photoperiodic photosensitivity, but also what has been termed 'internal coincidence'.

Colin Pittendrigh proposed that changing photoperiods may alter the internal phase relationships between two or more rhythms, bringing them together or apart either to initiate or inhibit photoperiodic induction. In Pittendrigh's model, light controls only the internal phase relationships between multiple circadian rhythms. These interactions in *Arabidopsis* are shown in Figure 23, and can be summarized as follows. The circadian clock regulates the expression and entrainment of two proteins: GIGANTEA (GI) and FLAVIN-BINDING KELCH REPEAT F-BOX 1 (FKF1). In the presence of blue light these two proteins form the GI-FKF1 complex which regulates photoperiodic flowering. Under long days (LD, see Figure 23), the peak expression of GI and FKF1 protein coincide (internal coincidence), and in the presence of blue light there is an accumulation of the GI-FKF1 complex. The GI-FKF1 complex regulates the degradation of rhythmically cycling DNA-binding one zinc finger (DOF) factors (CDFs). DOF proteins are members of an important family of plant transcription factors which bind directly to DNA. The CDFs bind to the promoter of another key gene which encodes the protein CONSTANS (CO) and inhibit its transcription. At dawn, CO transcription is repressed by CDFs bound to the CO promoter. But as GI and FKF1 accumulate after dawn and form the GI-FKF1 complex the CDFs are inhibited (external coincidence). This releases CO transcription from inhibition, leading to the accumulation of CO. The accumulation of CO protein then promotes flowering by binding, with co-factors and the regulatory regions of the FLOWERING LOCUS T (FT) gene. FT is a small mobile protein regarded as a 'florigen' or flowering hormone. The amount of FT influences the induction of flowering. It is synthesized from the leaf vasculature and transported to the shoot apical meristem to initiate flowering. Under short days (SD, see Figure 23), the rhythmic expression peaks of FKF1 and GI do not coincide. Without the accumulation of the light-dependent GI–FKF1 complex, CO expression is continuously suppressed by CDF proteins during the daytime, and no flowering signal is initiated.

The architecture of the photoperiodic control of flowering is even more complicated than outlined in Figure 23. But the original Bünning hypothesis, later modified by Pittendrigh, was remarkably

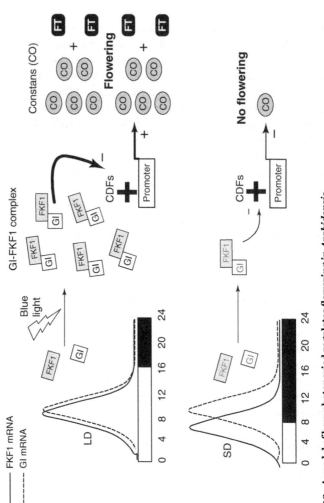

23. A model of how photoperiod regulates flowering in *Arabidopsis*.

prescient given that these ideas were all developed before the era of molecular genetics. Indeed, these concepts have provided a robust framework for investigating the cellular and molecular mechanisms underlying how a 24-hour circadian timing system can be used to generate seasonal biology in a broad range of organisms, including animals.

During the autumn and winter, the dual challenges of limited food availability and the need for additional energy to keep warm, especially among small animals, means that energetically demanding key life events, such as growth and reproduction, have to be timed to energetically favourable environmental conditions by some form of seasonal timer. Small mammals and birds in particular reduce the size of their reproductive organs and stop producing sperm or ovulating in the autumn and winter. For example, male and female hamsters are triggered to develop their reproductive organs when the photoperiod is longer than 12.5 hours, and then shut down the reproductive system after breeding. In large mammals such as sheep, the shortening days of autumn trigger reproduction, with conception in autumn and then birth of the lambs the following spring after a ~150-day gestation. The lambs are timed to arrive when there is plenty of fresh spring grass to sustain milk production in the ewe and then the weaning of the lambs. Arctic reindeer have multiple adaptations to help them cope with winter. Apart from developing thick coats with two layers of fur, the footpads shrink and tighten, exposing the rim of the hoof which cuts into the ice and crusted snow to stop the reindeer from slipping. These changes also enable the reindeer to dig down through the snow to find lichen to eat.

Animals below 5 kg that overwinter in harsh conditions may hibernate and survive through a significant decrease in energy expenditure. A true hibernator, like a chipmunk, can reduce its body temperature to nearly 0°C during hibernation and change its heart rate from 350 beats per minute to as low as four beats per minute. The heart rate of a bear also drops, though not as markedly, while body temperature remains nearly normal during this period. Bears are not true hibernators; their changes are more accurately called 'winter lethargy'. Animals that truly hibernate don't show biological sleep, except during the brief periods when they break hibernation for a day or so. By contrast, many winter lethargic species are able genuinely to sleep.

Although many birds may go into a state of 'torpor' in order to save energy to survive a cold night, the common poorwill, which is a small relative of the nightjar and found in western states of the USA, is the only bird which is known to hibernate throughout the winter, concealed in piles of rocks. The strategy of choice for most birds is migration.

About 65 per cent of all bird species migrate. In a process that can take weeks of preparation, migratory birds run through a repertoire of behavioural and physiological changes before they take off. A marked increase in appetite and food consumption begins about two to three weeks before migration. This is accompanied by an increase in the efficiency of fat production and storage. Readying for migration often involves a switch in diet. Many autumn migrants switch from insects to a diet of berries and other fruits that are high in carbohydrates and lipids. Most migratory species are solitary for much of the year but start to flock together before, or during, migration. This social behaviour seems to allow improved predator avoidance, food finding, and orientation. Another radical behavioural change is a shift from being active exclusively during the day to flying at night. This shift from diurnal to nocturnal behaviour occurs in many species during migration, including most shorebirds and songbirds.

Birds need to be ready to breed very shortly after they arrive in the northern breeding grounds. This means that the reproductive programme needs to be activated during the journey. Such activation of the gonads demands careful timing as the bird cannot be burdened with an increasingly heavy set of reproductive organs while it is still flying long distances. Yet it needs to have the appropriate reproductive behaviours turned on upon arrival to establish a territory and court a mate. For the return journey, when birds depart and fly south for the winter, their reproductive system has already regressed. The question is—how do birds know when to migrate or moult; or how do chipmunks know when to start hibernating; or monarch butterflies to migrate?

It is not local conditions such as temperature or rainfall. These factors may provide fine tuning, but not the overall drive. The driver is photoperiod. The exact day and time of migration may depend on whether there is a tail-wind, or a dry day, but these local factors are all secondary. The common thread is the change in photoperiod and a mechanism for registering changes in day length, and translating this signal into a neuroendocrine response.

Birds and mammals perceive and register the photoperiodic message in different ways. In mammals, light is perceived by the photosensitive retinal ganglion cells (pRGCs) of the eye. This information is relayed to the circadian clock within the SCN, which governs melatonin production by the pineal gland through a multi-synaptic sympathetic neuronal pathway (see Figure 19). Melatonin released by the pineal is the messenger of night length and hence the photoperiodic response in mammals. As the night length expands and contracts over the seasons, the duration of melatonin synthesis and release expands and contracts to mirror darkness.

By contrast, melatonin is not critical for the photoperiodic response in birds. Remarkably, day length in birds is detected directly by photoreceptors within the hypothalamus. Light passes through the skull and is perceived by several populations of 'deep-brain photoreceptors' (DBPs) that control seasonal reproduction. These include vertebrate ancient opsin, and possibly melanopsin which is the same photopigment of the mammalian pRGCs, but located within cells of the hypothalamus. It is also possible that another population of photoreceptors within the hypothalamus, based upon the photopigment OPN5, may contribute to photoperiodic light detection. Although the photoperiodic light-detecting mechanisms differ dramatically between birds and mammals, the downstream events seem to have been conserved and involve a circadian clock within a part of the pituitary gland called the pars tuberalis (PT), followed by the release of thyroid-stimulating hormone (TSH) from the PT and the activation or deactivation of thyroxin.

The photoperiodic signalling pathway illustrated in Figure 24 has emerged from recent detailed studies by UK and Japanese researchers. In mammals, the duration of darkness and hence melatonin release from the pineal gland is measured by a circadian clock within the PT of the pituitary gland. Summer-like melatonin signals (short nights so short duration melatonin release) activate a PT-expressed clock-regulated transcription regulator called EYA3 (EYA transcriptional coactivator and phosphatase 3), which promotes the synthesis of TSH. By contrast, birds use DBPs to detect dawn and dusk. But like mammals, the duration of light and dark is measured by a circadian clock within the PT, perhaps also activating a homologous transcription regulator like EYA3 (see Figure 24).

Although the light-detecting systems differ markedly between birds and mammals, the downstream photoperiodic mechanisms follow a similar course. Long days (LD) promote the synthesis and release of TSH from the PT. TSH then travels to and stimulates specialized ependymal cells in the basal hypothalamus called tanycytes to produce the enzyme deiodinase 2 (DIO2) which converts the prohormone of thyroxine (T4) into the active form of thyroxine—triiodothyronine (T3). Under short days (SD) the ependymal cells release only very low levels of TSH and under these conditions the enzyme deiodinase 3 (DIO3) is produced,

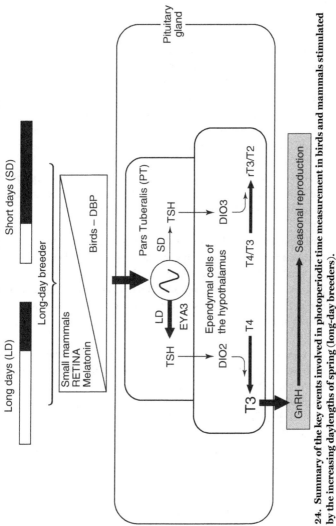

24. Summary of the key events involved in photoperiodic time measurement in birds and mammals stimulated by the increasing daylengths of spring (long-day breeders).

which degrades T4 and T3 into their inactive metabolites, reverse triiodothyronine (rT3) and diiodiothyronine (T2) (Figure 24).

Increased levels of T3 within the hypothalamus lead to the stimulation of gonadotropin-releasing hormone (GnRH) neurons to release gonadotropins. Gonadotropins promote the development of the reproductive organs, whose development increases the release of steroid hormones (testosterone, progesterone, oestrogen) (Figure 24). These stimulate brain receptors promoting reproductive behaviours such as song, territorial aggression, and courtship displays. T3 does not stimulate GnRH neurons directly but rather acts through distinct hypothalamic populations of neurons, which express two different peptides, namely kisspeptins (Kiss) and RFamide-related peptides (RFRPs), which are called gonadotropin-inhibitory hormone (GnIH) in birds.

Small mammals and birds are LD breeders, triggered by the increasing day lengths of spring, while sheep, deer, and other large mammals are SD breeders, triggered to breed in the autumn. As a result, DIO2 triggers reproduction in LD breeders but not SD breeders, while the action of DIO3 stimulates reproduction in SD breeders.

Photoperiod is the dominant source of information to time seasonal biology in species that live at mid-latitudes. Those living in environments where day length information is limited (e.g. at the equator where day length is close to 12 hours all year round, or in the high Arctic with its 24-hour summers and 24-hour winters), or have lifestyles that make day length unreliable or temporarily inaccessible (e.g. during migration and hibernation), use another timing system. These species rely on a clock with a period of about one year.

The hunt for these clocks started more than fifty years ago with the studies of a graduate student at the University of Toronto. Ted Pengelley asked a critical biological question: how is the behavioural physiology of an animal in hibernation, isolated underground, capable of emerging from hibernation at the right time and synchronizing the rest of its physiology with the seasons? To answer this question he kept golden-mantled ground squirrels (*Spermophilus lateralis*) in a light/dark cycle of 12 hours of light, 12 hours of dark (LD 12:12), and at a constant temperature in the laboratory, for up to nine years. Although the photoperiod did not change, the ground squirrels still entered and exited hibernation at about the same time each year. It was not exactly twelve months, and over time the circannual rhythm freeran with a period of about 10–11 months. Pengelley confirmed his original observations by also keeping ground squirrels in constant darkness or light. Collectively these studies were the first to prove conclusively that an internal circannual biological clock exists, ticking away under constant conditions and regulating repeated annual cycles of hibernation.

Working around the same time in Germany, Eberhard (Ebo) Gwinner noted an annual, spontaneous development of *zugunruhe* (night-time restlessness and agitation displayed by birds at the onset of migration) in captive willow warblers (*Phylloscopus trochilus*) held in constant conditions of light and dark. This provided the first unambiguous evidence for the involvement of an internal circannual clock in a seasonal timing event in birds.

Rhythms of *zugunruhe* and moult have been found to persist for at least ten cycles in many bird species kept under constant conditions—hidden from any photoperiodic signals. This is convincing evidence that a circannual timer can drive migration. However, most species seem to use both a photoperiodic/circadian timer and a circannual timer. In equatorial regions, the photoperiodic signal is weak, perhaps only a few minutes between day and night length across the seasons. This weak signal can be detected, but it is 'noisy' and augmented by a circannual timer to provide a robust signal to stimulate migration from equatorial to higher latitudes. In the far north, photoperiodic signals change very rapidly, and again birds use a rapidly changing and possibly difficult to detect photoperiodic signal combined with a circannual timer to time their migration from high to low latitudes. Sheep also rely on both photoperiodic and circannual timers to time their seasonal reproduction. Some species of sheep maintained under constant conditions of LD 12:12 show annual rhythms in gonadal development and regression. Such apparent 'redundancy' of circadian and circannual timing is a common feature in many physiological processes where different systems overlap to ensure physiological and behavioural precision.

Tracking down the location of the circannual clock has been difficult, but recent work has strongly suggested that the PT houses this timer. In key experiments, male European hamsters were deprived of seasonal time cues by pinealectomy (no melatonin signal) and maintained under a constant photoperiod. Hamsters still showed circannual rhythms in body weight and testicular size. Significantly, TSH expression in the PT showed a robust circannual variation with highest levels in the subjective summer state (Figure 24). This rhythm in TSH was correlated with rhythmic hypothalamic DIO2 and RFRP expression. Collectively the data suggested that TSH is a circannual output of the PT, which in turn regulates hypothalamic neurons controlling reproductive activity. The data also suggest that both the circannual and the photoperiodic signals converge on the PT to drive an annual pattern of TSH expression to synchronize seasonal biology to the seasons. Whether the PT contains the circannual timer of birds and other vertebrates remains to be determined.

Until the recent past, the changing seasons influenced human biology, with indications of annual cycles of reproduction, immune function, disease, and death. Since the electric light, air conditioning, central heating, and globalized food production, humans have become progressively isolated from seasonal changes in temperature, food, and photoperiod in the industrialized nations. Nevertheless, the seasons continue to have effects on our lives. The activity of almost a quarter of our genes (5,136 out of 22,822 genes examined) differs according to the time of year, with some more active in winter and others more active in summer. It is not just our genes—much of our biology, including the composition of our blood and fat tissue, changes across the seasons. While we do not know if these changes result from some photoperiodic and/or circannual timer, or are simply driven by the environment, it does raise the possibility that although seasonal rhythms in humans have been largely masked by social forces, they may yet turn out to be important contributors to our physiology and even disease.

Chapter 9
Evolution and another look at the clock

The circadian system controls every key aspect of biochemistry, physiology, and behaviour. All plants, fungi, protista, algae, invertebrates, vertebrates, cyanobacteria, and at least one archaeon display circadian rhythms. These rhythms appear to be evolutionarily ancient, with many molecular components conserved between diverse species, from *Drosophila* to mouse to human. Irrespective of taxa or species, in eukaryotes at least the rhythm is thought to be generated in broadly the same way, by molecular transcription–translation feedback loops (TTFLs) in which a group of core clock genes regulate each other to ensure that their mRNA levels oscillate with a period of ~24 hours (Chapter 5).

As circadian rhythms are found in every living thing, and as they seem to generate a rhythm in a broadly conserved manner, then it is not unreasonable to assume that circadian rhythms have been very useful, as they have been retained and refined by natural selection over billions of years. However, the empirical evidence that circadian rhythms contribute to the overall survival and reproductive success of an individual has been more difficult to demonstrate.

With constant food and water in a clean cage in the laboratory, rats and mice thrive. They also do perfectly well with an ablated

SCN and exhibiting no overt circadian rhythmicity. How such an animal would survive living freely in the wild, competing for food and avoiding predation, is another matter entirely. Patricia DeCoursey, at the University of South Carolina, attempted to answer this challenging question. As DeCoursey said:

> subjects must be located in pristine, preferably wilderness areas free of human interference, permanently marked, brought into the laboratory for the SCN-lesioning, then repatriated at their home dens within a few days. Subsequently, parameters of longevity, mortality, and daily activity rhythms, as well as specific physiological features must be tracked continuously for the subjects' remaining lifetimes. Automated telemetry such as radio tracking and implanted data loggers or transponders has eased the burden of these tasks, but the expense and energy demands for the tracker are still a major deterrent to data acquisition.

In a landmark experiment, DeCoursey and her students studied chipmunks for eighteen months in a four hectare plot of forest in the Allegheny Mountains. Seventy-four chipmunks were captured at the start and divided into three groups. All the animals were fitted with radio collars, and one group had their SCN lesioned and were arrhythmic, another group went through a sham surgical procedure without actual lesioning, and a third group were left as intact controls. The radio tracking collar could locate the animal to an area of approximately 0.5 metres in diameter, and movement could be detected by a continuously changing signal.

At the start of the study there was an extremely high population density of chipmunks due to favourable conditions in the previous two years. There were approximately 126 adult chipmunks resident on the project site, close to the maximum carrying capacity. Chipmunk mortality had been very low and until the start of the study no predators were detected. However, the abundance of chipmunks attracted a predator which captured predominantly SCN-lesioned chipmunks as prey. Within a few weeks the SCN-lesioned chipmunks were decimated in numbers. Weasels were the likely culprits, being ferocious predators that can eat their own body weight in prey each day. How the weasels detected the SCN-lesioned chipmunks is not clear, and no other large-scale study of this sort has been repeated. As a result we do not know specifically what made the SCN-lesioned chipmunks vulnerable.

Another approach, using more tractable species, has been to examine whether organisms are adversely affected when they are grown in a light/dark cycle with a period that does not match that of their endogenous circadian clock. The effect of light/dark cycle periods on growth rate in the cyanobacterium *Synechococcus elongatus* was studied by Carl Johnson at Vanderbilt University. Two strains of *S. elongatus* with different circadian period lengths were mixed and grown in culture flasks under light cycles of different periods (e.g. LD 10:10). The strain that had an endogenous period most closely matching that of the light cycle in which they were maintained rapidly out-competed the other strain and became the dominant strain in the culture flask. This convincing study showed that an organism seemed to benefit from having an endogenous oscillator that closely matches the environmental cycle. But again, the basis for this enhanced growth rate and increased fitness is unknown.

When *Arabidopsis thaliana* plants with mutations that result in altered period lengths were placed in an environment with light/dark cycles that were shorter, equal to, or longer than the endogenous circadian period length, those plants whose endogenous clock matched that of the external light/dark cycle had increased photosynthesis, growth, and survival. In similar studies of period length, *tau* mutant hamsters and *Drosophila* mutants that have non-24-hour circadian periods show significant (less than 20 per cent) reductions in lifespan when they are kept in 24-hour cycles. However, and once again, the mechanisms underpinning these deficiencies are not known.

Arabidopsis plants have an efficient defence response timed to coincide with the circadian-regulated feeding behaviour of the cabbage looper larvae (*Trichoplusia ni*). Cabbage loopers primarily feed during the day, and the plant's circadian system marshals a range of hormonal defences, particularly oxygenated fatty acid compounds called jasmonates, which are released by the leaves and reach peak levels at midday. Plants entrained in-phase with the loopers' feeding activity had visibly less tissue damage than plants entrained out of phase with the insects. The in-phase entrainment advantage is lost in plants deficient in jasmonate hormone. Jasmonates are released in response to wounding by insects and are responsible for inducing the expression of most wound-regulated genes in *Arabidopsis* leaves. However, it takes the plant time to synthesize the jasmonates. By employing a circadian regulator of jasmonate biosynthesis, anticipating the timed feeding behaviour of the loopers, the jasmonates can be released immediately after attack. This example provides a clear demonstration of how the possession of a circadian clock provides an advantage for the plant. However, the critical evidence (although likely) that this response increases the fitness (individual reproductive success) of the plant is lacking.

Although the evidence is less complete than one might have hoped, there is a consensus that the internal clock has been of vital importance in the evolutionary history of living things. One early and still persuasive idea, known as the 'flight from light' hypothesis, suggests that transcriptional oscillators may have provided a selective advantage early in evolutionary history by preventing the DNA-damaging effects of sunlight. Some three billion years ago, early life forms were subjected to intense ultraviolet bombardment during the day as there was no oxygen and hence no filtering ozone layer. Organisms that could partition DNA-based replication to the night by preparing in advance, rather than just responding to the presence of light, would be at an advantage.

Cyanobacteria have existed, in the form of stromatolites at least, for at least three billion years. Although they are unicellular with no nucleus, they time and synchronize many of their metabolic reactions with the environment. Indeed, while about 10–20 per cent of the genes of eukaryotes are rhythmically expressed, in cyanobacteria a much larger percentage of genes are under circadian control, with one study suggesting that all genes are under circadian regulation. Cyanobacteria separate photosynthesis (anabolism) during the day and respiration (catabolism) during the night. This temporal separation of largely incompatible metabolic processes appears to be one of the driving forces for evolving an endogenous timing mechanism. Three genes (*kaiA*, *kaiB*, and *kaiC*—'kai' means 'rotation' or 'cycle number' in Japanese) are essential components of the cyanobacterial circadian clock. At first, the cyanobacterial clockwork appeared to be just another TTFL in which clock proteins autoregulate the activity of their own promoters. However, although these proteins do interact to generate a TTFL, transcription and translation of these genes is not necessary for the generation of a circadian rhythm by the Kai proteins.

Takao Kondo of Nagoya University incubated the proteins KaiC, KaiA, KaiB, and adenosine triphosphate (ATP—the 'energy' molecule) in a test tube, and showed that just these four components were all that was needed to establish rhythmic phosphorylation and dephosphorylation cycles of KaiC. As illustrated in Figure 25, KaiA stimulates KaiC phosphorylation utilizing ATP, whereas KaiB induces KaiC dephosphorylation. This circadian rhythm of KaiC phosphorylation persists in constant darkness; it is temperature compensated, showing that this essential feature of a circadian oscillator must be embedded within the three Kai proteins and the nature of their interactions; and the phosphorylation of KaiC can be stably entrained to a temperature cycle, which in the wild would correspond to the light/dark cycle. In addition, KaiC appears to be part of an output pathway as it interacts with downstream proteins such SasA

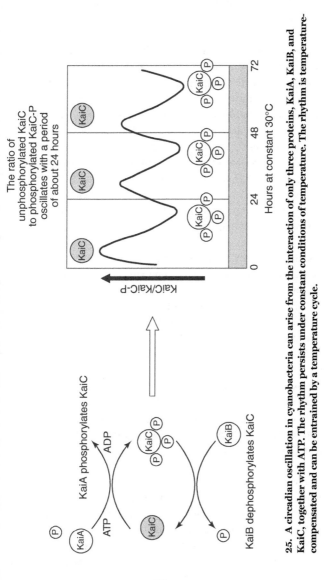

The ratio of unphosphorylated KaiC to phosphorylated KaiC-P oscillates with a period of about 24 hours

KaiA phosphorylates KaiC

KaiB dephosphorylates KaiC

25. A circadian oscillation in cyanobacteria can arise from the interaction of only three proteins, KaiA, KaiB, and KaiC, together with ATP. The rhythm persists under constant conditions of temperature. The rhythm is temperature-compensated and can be entrained by a temperature cycle.

to regulate circadian rhythms of nitrogen fixation, photosynthesis, and cell division.

These findings show that a biochemical 'post-translational oscillator' (PTO) operates as part of the molecular clockwork of cyanobacteria. However, although this PTO is sufficient for generating rhythms *in vitro* in a test tube, the cyanobacterial circadian system *in vivo* also generates a TTFL whereby the *kaiC* message codes for the regulatory KaiC protein, which not only suppresses the *kaiB* and *kaiC* genes, but also suppresses circadian expression of all genes in the cyanobacterial genome. It appears that the Kai-generated TTFL may amplify PTOs and is more robust and stable than the PTO, which in the wild would be important for dealing with unpredictable environmental changes.

The discovery of a PTO has changed the way we think about how circadian rhythms can be generated, and paved the way for yet more remarkable discoveries. Akhilesh Reddy and John O'Neill, at Cambridge University, have demonstrated circadian rhythms in isolated human red blood cells (RBCs). This was unexpected because mature RBCs lack a nucleus and organelles, and so do not possess any DNA and are therefore unable to synthesize any RNA. Nevertheless, they showed that the antioxidant protein peroxiredoxin (PRX) involved in scavenging intracellular hydrogen peroxide (H_2O_2), a by-product of a cell's normal energy-manufacturing activities, alternates rhythmically over approximately 24 hours between an oxidized and a reduced state.

Although the mechanisms underpinning the circadian PTO in RBCs remain to be clearly defined, these data, along with the cyanobacteria studies, raise the question. Perhaps most cells contain both a robust TTFL and multiple less stable biochemical PTOs? The role of a cellular TTFL would be to provide a very stable and entrained circadian rhythm that drives the transcription of key proteins, including clock proteins. In addition, another role of the TTFL would be to stabilize and amplify multiple, yet to be discovered biochemical PTOs. The cell-specific PTOs would then do much of the actual timekeeping by driving essential circadian rhythms in cellular biochemistry. This notion could explain why all the activities of a eukaryotic cell appear to be rhythmic, but only 10–20 per cent of the transcriptome is rhythmically expressed. Could the TTFL drive the transcriptome, while rhythmic biochemistry is driven by PTOs? Of course this whole notion may be nonsense. PTOs may turn out to be very rare and restricted to only a few specialized cells types. Either way, these new ideas about circadian organization are being explored and tested.

We will probably never know the real reason why circadian timing systems evolved, but what we do know is that a timing system of some sort seems to have been hardwired into the genome a very long time ago. The world in which clocks evolved is profoundly different from the one in which we live today. Even as recently as 150 years ago, before the advent of electrification, we lived by the patterns driven by the solar day. It has been said that we humans are prepared for a world that no longer exists. Our world lacks clear 24-hour patterns. The profound changes in light, temperature, and food availability in nature are all masked by our 24/7 society. With such ancient biological systems embedded within our biology, we are going to have to learn much more about how circadian systems work, develop, and control physiology and behaviour if we are to achieve health and productivity in our brave new 24/7 world.

Further reading

Chapter 1: Circadian rhythms: A 24-hour phenomenon

Dunlap, J. C., Loros, J. J., and DeCoursey, P. J. (2011). *Chronobiology*. Sinauer Associates Inc.

Foster, R. G. and Kreitzman, L. (2004). *Rhythms of Life: The Biological Clocks that Control the Daily Lives of Every Living Thing*. Profile Books.

Pittendrigh, C. S. (1993). Temporal Organization: Reflections of a Darwinian Clock-Watcher. *Annual Review of Physiology*, 55, 17–54.

Chapter 2: Time of day matters

Anderson, J. A., Campbell, K. L., Amer, T., Grady, C. L., and Hasher, L. (2014). Timing is everything: Age Differences in the Cognitive Control Network are Modulated by Time of Day. *Psychology & Aging*, 29, 648–57.

Biss, R. K., and Hasher, L. (2012). Happy as a Lark: Morning-type Younger and Older Adults are Higher in Positive Affect. *Emotion*, 12, 437–41.

Bollinger, T., and Schibler, U. (2014). Circadian Rhythms: From Genes to Physiology and Disease. *Swiss Medical Weekly*, July 2014.

Lemmer, B., Kern, R.-I., Nold, G., and Lohrer, H. (2002). Jet Lag in Athletes after Eastward and Westward Time-Zone Transition. *Chronobiology International*, 19, 743–64.

Thuna, E., Bjorvatn, B., Floc, E., Harris, A., and Pallesen, S. (2015). Sleep, Circadian Rhythms, and Athletic Performance. *Sleep Medicine Reviews*, 23, 1–9.

Chapter 3: When timing goes wrong

Arendt, J. (2010). Shift Work: Coping with the Biological Clock. *Occupational Medicine*, 60, 10–20.

Bhatti, P., Mirick, D. K., and Davis, S. (2012). Shift Work and Cancer. *American Journal of Epidemiology*, 176, 760–3.

Buxton, O. M., Cain, S. W., O'Connor, S. P., Porter, J. H., Duffy, J. F., Wang, W., Czeisler, C. A., and Shea, S. A. (2012). Adverse Metabolic Consequences in Humans of Prolonged Sleep Restriction Combined with Circadian Disruption. *Science Translational Medicine*, 4, 129ra43.

Chen, L., and Yang, G. (2015). Recent Advances in Circadian Rhythms in Cardiovascular System. *Frontiers in Pharmacology*, 6, 71–8.

Eisenstein, M. (2013). Stepping Out of Time. *Nature*, 497, S10–12.

Foster, R. G., Peirson, S. N., Wulff, K., Winnebeck, E., Vetter, C., and Roenneberg, T. (2013). Sleep and Circadian Rhythm Disruption in Social Jetlag and Mental Illness. *Progress in Molecular Biology and Translational Science*, 119, 325–46.

Jagannath, A., Peirson, S. N., and Foster, R. G. (2013). Sleep and Circadian Rhythm Disruption in Neuropsychiatric Illness. *Current Opinion in Neurobiology*, 23, 888–94.

Jones, C. R., Huang, A. L., Ptáček, L. J., and Fu, Y. H. (2013). Genetic Basis of Human Circadian Rhythm Disorders. *Experimental Neurology*, 243, 28–33.

Kondratova, A. K., and Kondratov, R. V. (2013). The Circadian Clock and Pathology of the Ageing Brain. *Nature Reviews Neuroscience*, 13, 325–35.

Marquié, J.-C., Tucker, P., Folkard, S., Gentil, C., and Ansiau, D. (2015). Chronic Effects of Shift Work on Cognition: Findings from the VISAT Longitudinal Study. *Occupational and Environmental Medicine*, 72, 258–64.

Pritchett, D., Wulff, K., Oliver, P. L., Bannerman, D. M., Davies, K. E., Harrison, P. J., Peirson, S. N., and Foster, R. G. (2012). Evaluating the Links between Schizophrenia and Sleep and Circadian Rhythm Disruption. *Journal of Neural Transmission*, 119, 1061–75.

Roenneberg, T., Allebrandt, K. V., Merrow, M., and Vetter, C. (2012). Social Jetlag and Obesity. *Current Biology*, 22, 939–43.

Chapter 4: Shedding light on the clock

Douglas, R., and Foster, R. G. (2015). The Eye: Organ of Space and Time. *Optician*, 20 March 2015.

Foster, R. G., and Hankins, M. W. (2007). Circadian Vision. *Current Biology*, 17, R746–51.

Foster, R. G., and Kreitzman, L. (2004). *Rhythms of Life: The Biological Clocks that Control the Daily Lives of Every Living Thing*. Profile Books.

Hughes, S., Jagannath, A., Rodgers, J., Hankins, M. W., Peirson, S. N., and Foster, R. G. (2016). Signalling by Melanopsin (OPN4) Expressing Photosensitive Retinal Ganglion Cells. *Eye*, 30, 247–54.

Jagannath, A., Butler, R., Godinho, S. I., Couch, Y., Brown, L. A., Vasudevan, S. R., Flanagan, K. C., Anthony, D., Churchill, G. C., Wood, M. J., Steiner, G., Ebeling, M., Hossbach, M., Wettstein, J. G., Duffield, G. E, Gatti, S., Hankins, M. W., Foster, R. G., and Peirson, S. N. (2013). The CRTC1-SLK1 Pathway Regulates Entrainment of the Circadian Clock. *Cell*, 154, 1100–11.

Moore, R. Y. (2013). The Suprachiasmatic Nucleus and the Circadian Timing System. *Progress in Molecular Biology and Translational Science*, 119, 1–28.

Ralph, M. R., Foster, R. G., Davis, F. C., and Menaker, M. (1990). Transplanted Suprachiasmatic Nucleus Determines Circadian Period. *Science*, 247, 975–8.

Schibler, U., Ripperger, J., and Brown, S. A. (2003). Peripheral Circadian Oscillators in Mammals: Time and Food. *Journal of Biological Rhythms*, 18, 250–60.

Welsh, D. K., Logothetis, D. E., Meister, M., and Reppert, S. M. (1995). Individual Neurons Dissociated from Rat Suprachiasmatic Nucleus Express Independently Phased Circadian Firing Rhythms. *Neuron*, 14, 697–706.

Yamazaki, S., Numano, R., Abe, M., Hida, A., Takahashi, R., Ueda, M., Block, G. D., Sakaki, Y., Menaker, M., and Tei, H. (2000). Resetting Central and Peripheral Circadian Oscillators in Transgenic Rats. *Science*, 288, 682–5.

Chapter 5: The tick-tock of the molecular clock

Aguilar-Arnal, L., and Sassone-Corsi, S. (2014). Chromatin Landscape and Circadian Dynamics: Spatial and Temporal Organization of Clock Transcription. *Proceedings of the National Academy of Sciences*, 122, 6863–70.

Hardin, P. E. (2011). Molecular Genetic Analysis of Circadian Timekeeping in *Drosophila*. *Advances in Genetics*, 74, 141–73.

Konopka, R. J., and Benzer, S. (1971). Clock Mutants of *Drosophila melanogaster*. *Proceedings of the National Academy of Sciences*, 68, 2112–16.

Partch, C. L., Green, C. B., and Takahashi, J. S. (2013). Molecular Architecture of the Mammalian Circadian Clock. *Trends in Cell Biology*, 24, 90–9.

Sehgal, A. (ed.) (2004). *Molecular Biology of Circadian Rhythms*. John Wiley and Sons.

Weiner, J. (2005). *Time, Love, Memory*. Vintage Books.

Chapter 6: Sleep: The most obvious 24-hour rhythm

Lockley, S. W., Foster, R. G. (2012). *Sleep: A Very Short Introduction*. Oxford University Press.

Saper, C. B., Scammell, T. E., and Lu, J. (2005). Hypothalamic Regulation of Sleep and Circadian Rhythms. *Nature*, 437, 1257–63.

Saper, C. B. (2013). The Neurobiology of Sleep. *Sleep Disorders*, 19, 19–31.

Chapter 7: Circadian rhythms and metabolism

Bass, J. (2012). Circadian Topology of Metabolism. *Nature*, 491, 348–56.

Jha, P. K., Challet, E., and Kalsbeek, A. (2015). Circadian Rhythms in Glucose and Lipid Metabolism in Nocturnal and Diurnal Mammals. *Molecular and Cellular Endocrinology*, 418, 74–88.

Kalsbeek, A., la Fleur, S., and Fliers, E. (2014). Circadian Control of Glucose Metabolism. *Molecular Metabolism*, 3, 372–83.

Chapter 8: Seasons of life

Dardente, H., Hazlerigg, D. G., and Ebling, F. J. P. (2014). Thyroid Hormone and Seasonal Rhythmicity. *Frontiers in Endocrinology*, 5, 1–11.

Foster, R. G., and Kreitzman, L. (2009). *Seasons of Life: The Biological Rhythms that Enable Living Things to Survive and Thrive*. Profile Books.

Greenham, K., and McClung, R. (2015). Integrating Circadian Dynamics with Physiological Processes in Plants. *Nature Reviews Genetics*, 16, 598–610.

Hut, R. A., Dardente, H., and Riede, S. J. (2014). Seasonal Timing: How Does a Hibernator Know When to Stop Hibernating? *Current Biology*, 24, R6025.

Song, Y. H., Ito, S., and Imaizumi, T. (2013). Flowering Time Regulation: Photoperiod- and Temperature-Sensing in Leaves. *Trends in Plant Science*, 18, 575–83.

Wood, S., and Loudon, A. (2014). Clocks for All Seasons: Unwinding the Roles and Mechanisms of Circadian and Interval Timers in the Hypothalamus and Pituitary. *Journal of Endocrinology*, 222, R39–R59.

Chapter 9: Evolution and another look at the clock

Egli, M., and Johnson, C. H. (2013). A Circadian Clock Nanomachine that Runs without Transcription or Translation. *Current Opinion in Neurobiology*, 23, 732–40.

Goodspeed, D., Chehab, E. W., Covington, M. F., and Braam, J. (2013). Circadian Control of Jasmonates and Salicylates: The Clock Role in Plant Defense. *Plant Signaling & Behavior*, 8, e23123.

Kondo, T., Strayer, C. A., Kulkarni, R. D., Taylor, W., Ishiura, M., Golden, S. S., and Johnson, C. H. (1993). Circadian Rhythms in Prokaryotes: Luciferase as a Reporter of Circadian Gene Expression in Cyanobacteria. *Proceedings of the National Academy of Sciences*, 90, 5672–6.

Ouyang, Y., Andersson, C. R., Kondo, T., Golden, S. S., Johnson, and C. H. (1998). Resonating Circadian Clocks Enhance Fitness in Cyanobacteria. *Proceedings of the National Academy of Sciences*, 95, 8660–4.

Reddy, A. B., and Rey, G. (2014). Metabolic and Nontranscriptional Circadian Clocks: Eukaryotes. *Annual Review of Biochemistry*, 83, 165–89.

Index

A

Acetyl-CoA carboxylase 104
acetylcholine 88
acrophase 4
action potentials 55, 56
adenosine A2 receptor 86
adenosine triphosphate (ATP)
    85, 126
adipose 57, 59, 92, 94, 95, 98, 99,
    101, 102, 104, 105
adrenocorticotropic hormone
    (ACTH) 94, 100
advanced sleep phase disorder
    (ASPD) 32
African clawed toad 51
Air India 25
Allard, Henry 108
Allegheny Mountains 123
alpha 46
alpha-melanocyte-stimulating
    hormone (alpha-MSH) 99
Alzheimer's disease 40, 41
aminergic neurons 86, 88
Androsthenes 7
angiosperms 109
anterior hypothalamus 33, 41
Aplysia californica 17
Aplysia fasciata 17

apolipoproteins 101
Arabidopsis thaliana 80, 110, 111,
    124, 125
Arctic reindeer 6, 113
arcuate nucleus 59, 98, 99
Aswan Dam 24
autonomic nervous system 56,
    98, 101

B

basal forebrain 41
bee 5, 6, 27
Benoit, Jacques 50
Benzer, Seymour 63, 64, 65, 66
Berson, David 51
beta cells 92
bHLH (basic helix-loop-helix)
    66, 67
bipolar disorder 37, 40
blood pressure 14, 26, 100
BMAL (Brain muscle arnt-like 1)
    73, 75, 77, 104
body temperature 26, 36, 58, 99,
    113, 114
Borbély, Alexander 84
Brandstaetter, Roland 22
Bünning, Erwin 8, 9, 109,
    110, 111

C

cabbage looper larvae 125
caffeine 19, 86
Calcium 60, 78
cAMP 78
cancer 23, 28, 29, 30
Carskadon, Mary 21
Casein Kinase 1 epsilon gene
    (*CK1e*) 73
CASEIN KINASE 2 (CK2) 69
cataracts 36
chipmunk 113, 115, 123, 124
cholesterol 22, 100, 101, 102
cholinergic neurons 86, 88
chromatin 79
chrononutrition 95
chronotherapy 22, 23
chronotype 17, 18, 19, 20, 22, 23,
    24, 90
circadian plasticity 6
circadian process (C) 84
circadian rhythm sleep
    disorders 32, 36
circadian rhythms 1, 2, 3, 4, 5, 6, 7,
    10, 11, 12, 14, 16, 18, 21, 26, 33,
    36, 40, 42, 44, 50, 53, 54, 55,
    56, 57, 58, 60, 63, 64, 77, 80,
    111, 122, 128, 129
circadian system 2, 3, 6, 9, 10, 14,
    20, 24, 25, 26, 28, 31, 36, 38,
    39, 42, 46, 48, 49, 52, 60, 61,
    79, 84, 86, 91, 97, 105, 113, 122,
    125, 126, 128, 129
circadian time 19, 46, 47
circannual clock 119, 120
Clk gene 71, 73
CLOCK (CLK) 67, 69, 71, 73, 75, 77
clock-controlled genes (*cgs*) 71,
    77, 80
clockwork orange (*cwo*) 71, 77
cognition 28, 33, 36
cognitive behavioural therapy
    (CBT) 44
colonic motility 95

common poorwill 114
cones 50, 51, 52
CONSTANS (CO) 111
core clock genes 2, 122
cortisol 26, 29, 56, 94
CREB-binding protein (CREB) 78
CREB-regulated transcription
    coactivator 1 (CRTC1) 78
Crick and Watson 62
CRY protein 69, 71, 73, 75, 77,
    102, 104
cryptochrome (*cry*) gene 69, 71,
    73, 75
CWO protein 71
cyanobacteria 3, 12, 80, 122,
    126, 128
CYCLE (CYC) 67, 69, 71, 73
CYP7A1 102
Czeisler, Charles 28

D

D site of albumin promoter
    (albumin D-box) binding
    Protein DBP 102
Dacey, Dennis 51
dark-loving (scotophilic)
    period 109
DBT (doubletime) kinase 69, 71,
    73
de Mairan, Jean Jacques Ortous 7
dead zone 47
Dec1 and *Dec2* 77
DEC1 and DEC2 77
DeCoursey, Patricia 123
deep-brain photoreceptors 115, 116
deiodinase 2 (DIO2) 116, 118, 120
delayed sleep phase disorder
    (DSPD) 33
depression 19, 21, 29, 37, 38,
    40, 42
diabetes 29, 30, 94, 97
diiodothyronine (T2) 118
DNA 36, 62, 64, 65, 66, 67, 79, 121,
    125, 128

DNA-binding one zinc finger
    (DOF) factors (CDFs) 111
Dodt, Eberhard 50
dopamine 41, 87
dorsomedial hypothalamus
    (DMH) 98
Dresslar, Fletcher Bacom 13
Drosophila 11, 17, 46, 49, 63, 64, 66,
    67, 72, 73, 75, 77, 79, 80, 122, 124
Dulles, John Foster 24, 25
Dunlap, Jay 80

E

E-box (Enhancer-box) 67, 69, 71,
    73, 75, 77
E cells 73
echidna 50
eclosion 11, 63, 64
ectothermic 60
endogenous clocks 3
entrain 2, 3, 4, 6, 9, 11, 38, 42,
    45, 46, 47, 49, 50, 52, 53,
    58, 59, 60, 72, 73, 75, 77, 78,
    79, 88, 89, 105, 109, 111,
    125, 128
epinephrine/adrenalin 94
ethyl methanesulfonate 64
eukaryotes 65, 122, 126
evening type 19, 22
external coincidence model 109,
    110, 111
extra-ocular photoreceptors 50
extra-retinal photoreceptors 52, 53
EYA3 (EYA transcriptional
    coactivator and phosphatase 3)
    116

F

fatty acid synthase 104
fibroblast 57
Feldman, Jerry 80
FLAVIN-BINDING KELCH
    REPEAT F-BOX 1 (FKF1) 111

flip flop switch 86
florigen 109, 111
FLOWERING LOCUS T (FT) 111
food-anticipatory activity
    (FAA) 58, 59
food-entrainable oscillator
    (FEO) 59, 60
Foster, Russell 50
freerun 4, 33, 41, 42, 45, 46, 47
Freerunning or non-24-hour sleep/
    wake rhythm 33
Frisch, Karl von 50
fruit fly 17, 49, 63

G

gamma-aminobutyric acid
    (GABA) 88
galanin 88
Galen 7
Garner, Wightman 108
gastric acid secretion 26, 95
ghrelin 29, 59, 98, 99, 100
GIGANTEA (GI) 111
glucagon 92
glucocorticoids 94, 100, 102, 105
gluconeogenesis 92, 94, 97, 99,
    100, 101
glucose 29, 91, 92, 94, 95, 97, 98,
    99, 100, 101, 102, 104, 105
glucose metabolism 12, 92, 98
glutamate 78, 88
glycogen 91, 92, 94, 97, 102
Golden, Susan 80
gonadotropin-releasing hormone
    (GnRH) 118
Great Circadian Disruption 31
growth hormone 94
Gwinner, Eberhard 119

H

Hall, Jeffery 66
Halobacterium 62
Hasher, Lynn 20

Harrison, John 10, 62
hibernation 113, 114, 118, 119
Hippocrates 7
histamine 86
Histone 79, 80
HLF (Hepatic leukaemia factor) 102
HMGCR reductase 22, 102
homeostatic 9, 84, 85, 86, 88, 97
homeostatic process (S) 84
homeotherm 60
Huntington's disease 41
hyperglycaemia 92, 94, 99
hypersomnia 36
hypoglycaemia 92
hypothalamus 50, 53, 56, 59, 97, 101, 104, 115, 116, 118
hyposomnia 36

I

insomnia 25, 33, 36
insulin 29, 92, 94, 97, 98, 101, 102, 104, 105
intermediolateral column (IML) 98
internal coincidence 110, 111
internal desynchrony 3, 25, 38, 79, 95, 105
irregular sleep timing 33

J

jasmonates 125
jet lag 3, 24, 25, 26, 27, 30, 78, 79, 89
JET protein 69
Johnson, Carl 80, 124

K

kaiA, kaiB, kaiC 126
Kay, Steve 80
kisspeptins 118
Kondo, Takao 80, 126

Konopka, Ron 63, 64, 65, 73
Kraepelin, Emil 38

L

larks 17
lateral hypothalamus 86, 88, 98
lateral neurons 66, 72
Lemmers, Bjorn 26, 27
leptin 29, 59, 98, 99
Lévi, Francis 23
Lewy body 41
lifestyle 30, 118
light-entrained oscillator (LEO) 60
light-loving (photophilic) period 109
light therapy 37
lipids 12, 91, 97, 100, 101, 102, 105, 114
locomotor rhythms 63, 64
long-day plants (LDPs) 108, 109
Longitude Prize 62
lovastatin 23
low-density lipoprotein 22

M

M cells 72
Mariana Trench 91
masking 47
meal times 44
melanopsin (OPN4) 51, 52, 116
melatonin 26, 42, 43, 88, 89, 98, 115, 116, 120
memory consolidation 3, 36, 37, 83
Menaker, Michael 50, 53, 58, 73
mental health 37, 39, 40
mental illness 26, 37, 40, 42, 43
mid-brain 41, 86
mid-sleep-time 17
mimosa 7
monoaminergic neurons 86, 88
Moore, Robert 53
morning types 20, 22

mRNA 3, 65, 66, 67, 71, 77, 78, 122
multiple sclerosis (MS) 42
Munich ChronoType
    Questionnaire, MCTQ 17
myocardial infarction (MI) 14, 16

N

Nasser, Colonel 24
neurodegeneration 36, 37
neuropeptide Y (NPY) 99
Neurospora crassa 80
night shift 28, 29, 30, 97
neuromedin S (NMS) 55
Nms cells 55, 56
nocturnin 101
noradrenaline 88
NREM sleep 88
nurses 28

O

obesity 29, 104
O'Neill, John 128
OPN5 116
optic chiasma 53
orexin neurons 86, 88, 99
owls 17

P

par domain protein 1 (pdp1) 71
paraventricular nucleus (PVN) 98,
    99, 100
Parkinson, James 41
Parkinson's disease 41
pars tuberalis 116, 120
PDP1 protein 71
Pengelley, Ted 118, 119
PER protein 66, 67, 69, 73, 75, 77,
    78, 102, 104
period (per) gene 64, 66, 67, 69,
    73, 75, 77, 78
peripheral clocks 58, 59, 92, 94,
    95, 97, 102, 105

peroxiredoxin 128
phase angle (*Phi* or *Φ*) 4
phase response curve (PRC) 46, 47
phase-shifting 45, 47, 60
Phylloscopus trochilus 119
photoperiod 6, 113, 115, 116, 119,
    120, 121
photosensitive retinal ganglion cell
    (pRGC) 51, 52, 55, 58, 77, 78,
    97, 115, 116
pineal 42, 50, 88, 98, 115, 115
Pittendrigh, Colin 11, 46, 64, 110, 111
pituitary 56, 78, 94, 100, 116
pituitary adenylate cyclise-activating
    polypeptide (PCAP) 78
platypus 50
positive feedback loop 37
post-translational oscillator
    (PTO) 128, 129
PPARs (peroxisome proliferator-
    activated receptors) 102
proopiomelanocortin (POMC) 99
prostaglandin 86
Provencio, Ignacio 51
pupa 63

Q

Q10 10

R

Reddy, Akhilesh 128
REM sleep 88
Reppert, Steven 55
retinohypothalamic tract (RHT) 97
REV-ERBα 77, 102, 104, 105
reverse triiodothyronine (rT3) 118
RFamide-related peptides
    (RFRPs) 118
Richter, Curt 8, 9, 53
rodless/coneless mouse 51
rods 50, 51, 52
Roenneberg, Till 17, 19
RORα 77

Rosbash, Michael 66
Rosenthal, Norman 37
runner bean 8

S

saliva 26, 95
schizophrenia 37, 38, 40
Schibler, Ueli 57, 58
SCRD 24, 28, 29, 30, 32, 36, 37,
    38, 39, 40, 42, 44, 100
seasonal affective disorder
    (SAD) 37, 42, 64
serotonin 37, 88
SHAGGY (SGG) kinase 69
shift work 3, 19, 24, 28, 29, 30,
    97, 105
short-day plants (SDPs) 108, 110
simvastatin 23
Silver, Rae 56
Sirtuin 1 (SIRT1) 104
sleep 3, 8, 17, 18, 19, 20, 21, 22, 24,
    28, 29, 32, 33, 36, 37, 41, 44,
    49, 58, 59, 60, 80, 81, 82, 83,
    84, 85, 86, 88, 89, 91, 97, 99,
    100, 104, 105, 114
sleep gate 84
sleep pressure 32, 84, 85
sleep reservoir 32
sleep stabilization 40, 41
sleep/wake 8, 12, 32, 33, 38, 41, 42,
    43, 60, 84, 86, 89, 91, 100
SLK1 78, 79
social jet lag 19, 20, 21
Spermophilus lateralis 119
statins 22, 23
stress axis 38, 39
substantia nigra 41
suprachiasmatic nucleus (SCN) 53,
    54, 55, 56, 57, 58, 59, 60, 61,
    77, 78, 79, 85, 88, 92, 94, 95,
    97, 98, 99, 100, 102, 105, 115,
    123, 124
Synechococcus elongatus 80, 124

T

Takahashi, Joe 55, 73
tamarind 7
tanycytes 116
tasimelteon 42
Tau 4, 53, 73, 124
TEF (thyrotroph embryonic
    factor) 102
teleost multiple tissue opsin
    (TMT) 50
temperature compensation 10, 11, 72
thalamus 55
thyroid-stimulating hormone
    (TSH) 116, 120
thyroxine 94, 116
TIM protein 67, 69, 71, 72, 75
Time Ball 45
timeless gene (*tim*) 67, 69
torpor 114
transcription–translation
    feedback loops (TTFLs)
    65, 67, 71, 75, 77, 80, 102,
    122, 126, 128, 129
trans-meridian flight 28, 30
triiodothyronine (T3) 116
triglycerides 92, 100, 101

V

Van Cauter, Eve 29
vasopressin (VP) 57
ventrolateral preoptic nuclei
    (VLPO) 88
ventromedial hypothalamus
    (VMH) 98
VRI protein 71
vrille (*vri*) 71

W

wake/sleep 32
Wauchoipe, Robert 45

Whitmore, David 50
World Health Organisation 29

Y

Young, Michael 66

Z

zebrafish 50
zeitgebers 2, 4, 6, 9, 11, 60, 61
Zucker, Irving 53
zugunruhe 119

Index

SOCIAL MEDIA
Very Short Introduction

Join our community

www.oup.com/vsi

- Join us online at the official Very Short Introductions **Facebook** page.
- Access the thoughts and musings of our authors with our online **blog**.
- Sign up for our monthly **e-newsletter** to receive information on all new titles publishing that month.
- Browse the full range of Very Short Introductions online.
- Read **extracts** from the Introductions for free.
- If you are a teacher or lecturer you can order inspection copies quickly and simply via our website.

ONLINE
CATALOGUE
A Very Short Introduction

Our online catalogue is designed to make it easy to find your ideal Very Short Introduction. View the entire collection by subject area, watch author videos, read sample chapters, and download reading guides.

http://global.oup.com/uk/academic/general/vsi_list/

GALAXIES
A Very Short Introduction
John Gribbin

Galaxies are the building blocks of the Universe: standing like islands in space, each is made up of many hundreds of millions of stars in which the chemical elements are made, around which planets form, and where on at least one of those planets intelligent life has emerged. In this *Very Short Introduction*, renowned science writer John Gribbin describes the extraordinary things that astronomers are learning about galaxies, and explains how this can shed light on the origins and structure of the Universe.

www.oup.com/vsi

SLEEP
A Very Short Introduction
Russell G. Foster & Steven W. Lockley

Why do we need sleep? What happens when we don't get enough? From the biology and psychology of sleep and the history of sleep in science, art, and literature; to the impact of a 24/7 society and the role of society in causing sleep disruption, this *Very Short Introduction* addresses the biological and psychological aspects of sleep, providing a basic understanding of what sleep is and how it is measured, looking at sleep through the human lifespan and the causes and consequences of major sleep disorders. Russell G. Foster and Steven W. Lockley go on to consider the impact of modern society, examining the relationship between sleep and work hours, and the impact of our modern lifestyle.

www.oup.com/vsi

CANCER
A Very Short Introduction
Nick James

Cancer research is a major economic activity. There are constant improvements in treatment techniques that result in better cure rates and increased quality and quantity of life for those with the disease, yet stories of breakthroughs in a cure for cancer are often in the media. In this *Very Short Introduction* Nick James, founder of the CancerHelp UK website, examines the trends in diagnosis and treatment of the disease, as well as its economic consequences. Asking what cancer is and what causes it, he considers issues surrounding expensive drug development, what can be done to reduce the risk of developing cancer, and the use of complementary and alternative therapies.